AF574641

de Courtivron's

AN IDENTIFICATION & VALUE GUIDE

Gael de Courtivron

with Lara de Courtivron

COLLECTOR BOOKS

A Division of Schroeder Publishing Co., Inc.

This book is dedicated to

Marguerite Behoteguy de Téramond

for all the years of love and patience

The current values in this book should be used only as a guide. They are not intended to set prices, which vary from one section of the country to another. Auction prices as well as dealer prices vary greatly and are affected by condition as well as demand. Neither the Author nor the Publisher assumes responsibility for any losses that might be incurred as a result of consulting this guide.

CREDITS

Research and Assistant Editor
Lara de Courtivron

Production and Layout
Gael de Courtivron
Robert E. Demperio
Lara de Courtivron
Karen Geary
Beth Summers

Assistant Art Direction
Robert E. Demperio

Photography
Randall E. Tosch
Bill Beaudin
Ugo Fadini
Gael de Courtivron

ON THE COVER

Top left: Buddy "L" #5117, c. 1970.
In circle: Marx #1090, c. 1957.
Bottom right: Metalcraft, Art Deco, c. 1934.

Additional copies of this book may be ordered from:
COLLECTOR BOOKS
P.O. Box 3009
Paducah, Kentucky 42002–3009

@ $24.95. Add $2.00 for postage and handling.

ACKNOWLEDGMENTS

Writing a book like this would not be possible without the efforts and support of many people. I am thankful to so many good friends for providing me with pictures and information, as well as their time and knowledge. The following people have demonstrated that there is a true fellowship between Coca-Cola toy truck collectors.

Al Wilson, Lyle Dever, Bob Spahr, Dale Kelley, Barry Brooks, Jim Kennedy, Harold Ford, Don Slayton, Marc Cardelli, Daniel Tuffery, Richard Catapane

With Special Thanks to...

The Coca-Cola Company/Phillip F. Mooney and Joanne Newman
For making themselves and the archives available to me for my research. Their support is much appreciated.

Lara Michelle de Courtivron
My daughter, for countless hours of research, writing, and editing as my assistant in this project. This book could not have been possible without your help. I Love You!

Allan Petretti
My friend and colleague for sharing his experience in writing books with me, and for spending a lot of his time advising me on this project. His advice and knowledge is much appreciated.

Ugo Fadini and Maria Teresa Biasio
My very dear friends and colleagues from Padova, Italy. Their help with the European chapter was essential, as were the many beautiful photos of their collection, when they were "needed yesterday."

Fred Thompson
For many interviews, and sharing of his invaluable knowledge on the history of Smith-Miller, as well as providing some great photographs from the Smith-Miller warehouse.

Kyle Foreman
My good friend and colleague, for loaning many trucks from his legendary collection for me to photograph as well as for his undaunting support for this project since day one.

John Varga
A good friend to collectors everywhere. The first Coca-Cola truck collector I ever met, John encouraged me from the very beginning. Also, for lending some beautiful trucks from his collection for me to photograph, and for always being available to share his knowledge on Coca-Cola trucks.

Dave Zorn
A good friend who is always willing to help out. For being one of my biggest supporters for this project. Dave's enthusiasm is contagious.

Randy Schaeffer and Bill Bateman
The C.C. Tray-ders, for always being the definitive source of information when I got stumped and, more specifically, for preparing the text on the American Flyer Coca-Cola Train, as well as for lending me some vital research material and for providing me with photographs. Their knowledge of the history of Coca-Cola is unparalleled.

Roy and Janeen Davis
Their diligent effort in photographing some of the rarest and most beautiful trucks from their extensive truck collection is much appreciated. They went out of their way to facilitate this project, truly exemplifying the spirit of collecting.

Ed and Sandy Campbell
For also giving me access to their vast collection, and for providing me the opportunity to photograph some of the rarest toys I have ever seen. Their collection is *spectacular!*

Carter Davis
For providing photos and always making himself and his incredible collection available to me.

Forrest Wright
For lending me one of his rarest and most prized "babies" to photograph. I truly appreciate the trust and friendship.

Ray and Debbie Kozlowski
My good friends! Who let me come to their house and systematically clear their showcases when I needed to photograph their trucks. Their constant support for this project was greatly appreciated.

Tanja Goinar and Detlef Siegel
For their help with the German section of the book, and providing and translating great research materials.

Rudy LeCoadic
The beautiful photographs he provided of his Göso are much appreciated.

Richard McNary
For sharing his vast knowledge on Marx toys, and providing toy catalogs for me to photograph. And especially for selling the Deco Metalcraft to me in 1985.

Stan Swartz
For sharing his great expertise on Matchbox and Lledo, his help was much appreciated.

The Marx Company – Miami, Florida
For granting me research access to their historical archives.

K.K."Butts" Makela
For constant support and motivation for this project from the beginning.

Anthony "Snake" Labua
I'm not sure he deserves thanks for this, but as one of my oldest and dearest friends, Tony started this whole process in 1978 when he educated me on the wonders of Coca-Cola collecting, and showed me my first Coke truck. Yeah, what the heck...thanks, my friend!

Rosalie C. (Shugs)
For a lifetime of putting up with my quirks and for letting me pursue my dreams. I love you.

IN MEMORIAM

The following collectors, although no longer with us, played an important role in my development as a Coca-Cola toy truck collector. I miss all of them dearly!
"Pop" Poppenheimer, Dave and Melba Caldwell, Bill Petrick, and Peter Ottenheimer.

CONTENTS

The Author

Gael de Courtivron, born and raised in Paris, France, has had the rare opportunity to live in many exotic places. The son of a retired foreign service officer, Gael was a "government brat" who has enjoyed all the perks of a multi-cultural life. Gael, a veteran rock 'n' roll drummer and vocalist, retired after 20 years in the music industry and has become an avid collector of Coca-Cola memorabilia for the last 15 years. Although Gael has a very diverse collection, he quickly developed a special affinity for the Coca-Cola toys. He has one of the most extensive Coca-Cola toy collections anywhere. He has been a writer/consultant to several Coca-Cola price guides and publications, and has also served as the Coca-Cola advisor to *Schroeder's Antique Price Guide* for the past 10 years.

Gael often operates under the name of Cocaholics, a company he formed 10 years ago, solely dedicated to the buying, selling, and trading of Coca-Cola memorabilia on an international basis. In 1988, he formed another umbrella company called Classics by Cocaholics in which he specialized in the restoration and sale of vintage Coca-Cola machines for such commercial accounts as the Coca-Cola Company, McDonalds, and various commercial designers, as well as for private collectors and savvy investors. Gael has traveled extensively, serving clients from Europe to Australia, as well as the United States. Gael also appraises Coca-Cola collectible items.

In 1991, Gael dissolved Classics by Cocaholics to pursue his dream of designing nostalgia products and in 1992 formed his newest and now sole venture, GAELART Concepts, Inc., in Sarasota, Florida. GAELART is a company specializing in the design and development of licensed Coca-Cola Brand™ merchandise. Through a deep and rich experience with Coca-Cola products, memorabilia and his creative ability, he successfully designs Coca-Cola Brand™ merchandise that has produced high volume sales. He currently resides in Sarasota, Florida, with his family.

You've probably all heard these words at one time or another…"It's too late," "You can't find them any more," "They're all in collections," or maybe even "what's the use." I know that I've heard those words…I guess these must be some of the ones that "weren't out there." Good luck!

– Gael, 1994

PREFACE

As John Styth Pemberton was mixing his legendary elixir in 1886 Atlanta, no one realized that it would ultimately lead to the creation of a company, which by its advertising alone would be responsible for such a phenomenon 100 years into the future. Today, Coca-Cola collectors are as ardent and faithful a group as that found in any other collectible field, if not more. The sheer beauty and appeal of the advertising pieces used to promote the company is the reason why they are sought after the world over. One of the great pleasures derived from collecting the memorabilia of Coca-Cola is the chance it has given us to observe the changing face of society through the years. Through Coca-Cola advertising we get a rare opportunity for a retrospective view of our lives, as well as those of our parents and grandparents, as they have evolved over the decades since those early days in Victorian Georgia.

Unlike any other single collectible, there are in fact many distinct areas within the Coca-Cola collecting field itself. Some Coke collectors are partial to trays or maybe calendars, others prefer bottles and cans, while still others may specialize in signs or maybe ads; nevertheless there is one area of this fascinating realm that always seems to put a smile on a face, Coca-Cola toy trucks. Could this be because we all secretly want to be young again, and these little toys bring out the kid in us? Since we're adults we try to keep quiet about our secret desires to roll these trucks on the floor, imitating the sounds of engine roars and brake squeals. We have personally known some very respectable members of society who have admitted to these little quirks, but all quirks aside, whether in the closet or not, we are beyond being Coke collectors, we have also assumed the role of guardians of these memories from a bygone era; when these impressive yellow and red machines roamed every street of the city and hamlet alike delivering smiles. We are, in fact, keepers of a little piece of history.

Today, Coca-Cola toy trucks are of multi-collectible value; not only are they Coca-Cola, but they are also toys, as well as specific toys manufactured by a specific company, in itself collectible. There could be even more brand name collectors than there are Coca-Cola collectors. Collectors who specialize in toys manufactured by companies like Smith-Miller, Buddy "L," Marx, Göso, Dinky Toy, Matchbox, or Marusan, to name a few. This is another reason for their wide-spread appeal. Thus, we identified a real need to create a book like this; one that would bring photographs and information to all collectors, whether they are advanced or just starting out – today or in the future – a book that could let everyone enjoy these wonderful Coca-Cola toy trucks.

A Metalcraft miniature truck window display installed in a Cincinnati department store. *The Red Barrel*, May 15, 1933.

This book is the most complete guide ever assembled solely dedicated to these marvelous Coca-Cola toy trucks, their identification and their current values. It encompasses most trucks produced between 1930 and 1986, with few exceptions. Although we strived to include all of them, we probably missed a few. As a matter of fact, we are aware of several that we wanted to include but were unable to convince their owners to provide us with photographs. This was the very rare exception, however, as so many collectors literally went out of their way to help us with this project. We are grateful to our friends for sharing their time, knowledge, and their collections with us; especially to those who supplied us with photographs and in many instances, trusted us enough to actually send their prized possessions to us, to be photographed.

Since 1986 so many collector type toy trucks have flooded the market that, due to space constraints, we have chosen that date as a cut-off point. So, if you know of some trucks that you feel were omitted, we welcome your suggestions and comments. Let us know and we will try to include them in our future book, *Coca-Cola Toys, Banks, Games, and Miniatures.*

To better appreciate and understand this book, it is important to have an historical perspective on these toys.

In the late 19th and the beginning of the 20th century, metal toys were mostly made of cast-iron, but due in part to excessive flaws, a high scrap rate, inaccurate detailing, and the fact that they were simply too heavy for kids, the toy industry searched for new materials to produce toys. By 1910, steel and tin toys were beginning to phase out these klunky cast-iron dinosaurs. Although produced concurrently for over a decade, stamped steel became the focus of the industry in the United States, while the Europeans focused more on tin. The advent of die-casting, stamped steel, tin lithography, and plastic injection continued to modernize the industry.

During the course of reading this book you will see certain terminology that you may have heard before but were unsure of exactly what it meant. We will try to clarify a few of these more common terms for you now.

Stamped or Pressed Steel: Although two different names, the process is the same, while some experts feel that pressed is actually a misnomer. The stamped steel process simplified is as follows: blank parts were stamped out of metal and formed into shape using tools and dies that were carefully made to ensure uniformity for quick assembly. Once the various parts were shaped, they went through a rattle mill to remove any of the sharp edges that could lead to injury. Then on to the spot welding department. From this department a few other bench operations were executed, and the trucks, still in several sections, were then dipped in enamel paint and baked at a very high temperature, usually in excess of 400°F, for a couple of hours. The trucks then went to the assembly line where they were put together, using rivets or screws, and completed by the addition of wheels, other accessories and decals when applicable. The most famous examples of Coca-Cola stamped steel trucks include Metalcraft, Buddy "L," and the early Marx Stake trucks.

Although toys produced as early as the 1800s were made from tin, it is important not to confuse a tin toy with a tin lithographed toy. The early tin toys were hand-painted. The lithographic process came about later as technology and machinery became available. Although commonly referred to as tin, the metal used in the tin lithography process was actually a sheet of steel with a tin-plated finish.

Tin Lithography is the process by which a design is painted onto a flat metal sheet. The metal sheet was first run through a machine where it was coated to ensure that colors would not be absorbed into the metal and blurred. The next step took it to the actual lithography machine. This machine had four individual rollers – two in front to mix the ink; one on top, the composition roller; and one on the bottom made of steel, with a blade that kept the bottom side of the metal sheet dry. A glass litho plate, which had the design on it, was placed into the lithography machine so that the inks would travel through the front rollers and onto the plate. There were four colors used in the process – red, yellow, black, and blue. A process called overlapping made it possible for two colors, overlapped, to produce a third color. Through this process, the original four lithographed colors could produce as many as nine different colors. Using a screening technique based on the number of dots per square inch resulted in the ability to control the depth of the color. Unlike their stamped steel cousins, tin lithographed toys needed tabs instead of screws to be assembled. The baking process was also much shorter, usually 10–12 minutes in large ovens set at 250-350°F. A machine could lithograph several dozen sheets of steel per minute, making it a very cost-effective process. The most popular tin lithographed Coca-Cola toy trucks were produced in Japan throughout the 1950s and 1960s; however, the highest quality models were produced in Germany by Göso and Tipp & Co., and in the United States by Louis Marx.

Still another process is called **Molded or Injected Plastic**. Molten colored plastic is poured into a mold, allowed to cool, and released creating a toy. Even though these molds were not excessively large, they were so heavy that they could only be moved by forklift. Some of the more popular Coke trucks made from this process were manufactured by Pyro and Marx. For more information on plastics see Marx in Chapter 4, page 41.

The **Die-cast** process is similar to molded plastic and is just what the name implies, casting a part by pouring molten metal into a die. The metal is then cooled and a casted shape is extruded. Die-casting can be accomplished using a variety of metals ranging from aluminum to zinc. This process originated with the linotype machine in 1893. Type casting, as it was originally called, was used to produce small premiums, political items, and penny jewelry. The first die-cast toy truck was produced by Tootsietoys in the United States, c.1910. Between 1933 and 1979, Dinky Toys enjoyed being the world leader in die-cast vehicles including the Coca-Cola Lorrey produced in the mid-1960s. Perhaps the most desirable die-cast Coke trucks, however, were produced in the United States by Smith-Miller. Other famous die-cast Coca-Cola toys were manufactured by the Mettoy Company, later called Corgi, and Lesney with its popular #37 Matchbox Coke truck and Y–12 Yesteryear series.

Finally, a quick note on **accessories**. Many of these Coca-Cola toy trucks came with accessories, therefore we have tried to list all of the accessories that were included with each particular truck, ranging from the 10 miniature glass bottles found on the Metalcraft trucks to those little Coca-Cola cases which were available in a myriad of colors, shapes, sizes, and materials – some with fixed and some with removeable bottles. And dozens of different hand trucks, miniature Coke machines, conveyor belts, forklifts, and little delivery people have made kids happy for nearly a century as they played with their favorite beverage truck. The accessories that have survived over the years greatly increase the value of these wonderful toys. We hope that you will enjoy this book as much as we have enjoyed writing it.

HAPPY COLLECTING!

Some of the accessories that came with the beautiful Coca-Cola trucks featured in this anthology.

HOW TO USE THIS GUIDE

The hardest part about writing a book like this is not simply assembling several hundred pictures between two covers but, rather, creating a useful guide for collectors everywhere. "Guide" is the key word here because, as the term implies, we have tried to create just that, a guide for the identification and value of Coca-Cola toy trucks. Please remember that this is only a guide. The actual value on any truck is determined by the end result of the deal between buyer and seller.

RARITY SCALE

We have assigned every truck in this book a number between 1 and 5, forming a "Rarity Scale." The most common trucks are represented by the number 1. These examples were mass produced and most collectors already have these trucks. On the opposite end of this scale is the number 5 for the rarest trucks. Often times these are trucks that we have only seen once or twice in the years that we have been collecting. These are the most desirable. The other trucks fall somewhere in between. Please refer to this scale to understand the price range.

PRICE RANGE

Unlike most guides, we have elected to use a price range rather than assign a fixed value to any truck. Usually, we never see identical prices on identical trucks, as prices seem to vary greatly. Unlike a supermarket or retail store, prices depend greatly on a variety of factors such as who is selling, the geographical area, the knowledge of the seller, and most importantly – how badly the buyer wants a certain truck. Due to these variables it wouldn't be a good barometer to quote the gift price that some lucky collector may have paid at a neighborhood garage sale. Consequently, it wouldn't be any more realistic to quote the price realized at auction for a particular truck that a very serious collector would pay. We have, therefore, chosen to use a realistic price range system. As a guide, we have used our personal knowledge of the market, studied auction prices versus show prices, and have taken into consideration the rarity and desirability of each truck to formulate this range. Keep in mind, however, that they are only approximate values and that these can and will vary greatly even beyond this range. In addition, there is yet another and probably *the most important factor* to consider when determining the value of a truck – condition.

UNDERSTANDING CONDITION

It is very important to understand condition when using a book like this. The prices quoted in this book reflect trucks in excellent condition, which typically means a truck in 8– 8.5 condition (10 being perfect). To be in this condition, a truck would have to have nearly all of its original accessories, only very slight pitting or chipping and minimal decal wear. This is our basis for the middle of our price range. For example, a truck ranging between $300.00–500.00 in 8.5 condition could actually be worth only $175.00 if it is dented, the cases are missing, the decals are worn off on the back, and it has many chips in the paint. On the other hand, if this same truck is in flawless condition, packaged in its original box with all the accessories still wrapped, and the original color brochure included, it could certainly be worth $750.00 or more. This shows exactly how much condition can affect the value of a particular truck.

REFERENCE NUMBERS

In order to simplify the identification of the trucks in this book, whether it be for insurance purposes, sales, or for simple reference, we have designated a reference number to each truck. You will find this number incorporated at the end of each truck's statistical data.

METALCRAFT

1931–1935

METALCRAFT

Metalcraft trucks are among the most collectible and desirable toy trucks on the market today. The Metalcraft Corporation of St. Louis, Missouri, began manufacturing these toys in the late 1920s (c.1929), and continued until the mid-1930s. However, the first Metalcraft trucks did not actually hit the market until February of 1931 when introduced at the New York Toy Fair. The person responsible for Metalcraft's success was Alfred Korte, the chief engineer and factory manager, who invented and designed everything Metalcraft produced.

Metalcraft trucks were made of stamped steel and offered a much needed alternative to the toy market dominated by cast-iron toys, which were beyond the economic means of many Americans during the Great Depression. What Metalcraft offered was a good quality, durable toy at a very affordable price that typically ranged from 29¢ to 49¢. Unlike cast iron, Metalcraft trucks were made from 20-gauge stamped steel – the same material as real cars and trucks.

The first Metalcraft truck to display any advertising aside from the actual word "Metalcraft" was the famous Coca-Cola bottling truck introduced in the fall of 1931, erroneously dated 1927. This was about six months after the New York Toy Fair where Metalcraft was first introduced. In 1931, Metalcraft's Coca-Cola truck was the only bottling toy truck in production.

There is speculation as to why the trademark "Coca-Cola" appeared on Metalcraft's first advertising truck. We believe that the connection is to be found with Robert Woodruff, the most influential man in Coca-Cola's history. Before becoming Coca-Cola's president in 1923, Woodruff had been president of the White Motor Company. In fact, Woodruff concurrently served as both president of White and Coca-Cola throughout 1923, commuting between Cleveland, Ohio, and Atlanta, Georgia, for the entire year. It was Woodruff who replaced the Coca-Cola company's horsedrawn wagons with an entire fleet of White trucks. This leads us back to the Metalcraft Corporation, which equipped their earliest Metalcraft trucks with a cadmium-plated, flat radiator plate with the name White embossed on it. The White Motor Company and Coca-Cola were in reality very much intertwined.

As soon as it hit the market, the little red and yellow Coca-Cola bottling truck, with 10 miniature bottles, was an immediate success and became the #1 toy on every child's Christmas wish list to Santa. It was so popular in fact that between the St. Louis Coca-Cola Bottling Company's premium store and other area department stores, over 35,000 trucks were sold during the pre-holiday toy season of 1931. Several other American and Canadian cities experienced the same phenomenon.

Observing the success of the Coca-Cola bottling truck, several new companies asked the Metalcraft Corporation to produce trucks advertising their products. Among these companies were Best Foods Mayonnaise, Heinz 57, Kroger, Shell Oil, Sunshine Biscuits, and Goodrich Tires. Using these trucks as promotional items for commercial businesses was an instrumental factor allowing Metalcraft to keep prices down by selling large quantities of trucks to their clients.

The Metalcraft Corporation produced four distinct versions of the Coca-Cola bottling truck. The earliest truck had stamped and polished metal wheels, a windshield post that was separated from the cab roof, and a

White embossed grille with no bumper. The second version saw the introduction of Goodrich rubber tires. This model had a windshield post and cab roof that were connected to make one piece, an added bumper, and the White grille was used again. Substantial change came with the third Metalcraft version of the Coca-Cola truck which now had battery-operated working headlights. The White grille was replaced with a removable one-piece grille and bumper to facilitate access to the battery cavity, and a switch was added to the right side of the cab to activate the headlights. In 1934, Metalcraft sent a memo to all Coca-Cola bottlers touting the arrival of their "newest and sleekest" model yet. (See page 18.) This fourth and final version, is often referred to as the "Art Deco" truck. This was Al Korte's greatest engineering triumph. The new streamlined cab was a pure work of art and remains the flagship of most collections. The "Art Deco" was also battery operated and sported yet another non-removable grille variation. This truck today remains the rarest and most sought after of the series.

Despite their variations in cab styling and wheel design, the four Coca-Cola trucks produced by Metalcraft were similar. Every truck shared the same yellow A-frame bottle carrier with the Coca-Cola trademark on the top of the bottle rack and rear of the truck, while the rest of the truck was red. All trucks came with ten individual miniature Coke bottles which were produced by two separate manufacturers and came in three different versions. The manufacturers were Chattanooga Glass, identifiable by a "C" within a circle, and Owens Illinois, identifiable by an "I" overprinted over an "O" and surrounded by a diamond. These two trademarks appear on the bottom of every miniature Coke bottle. Throughout our research, we have come to believe that the first bottles used were actually filled with brown liquid since

we have found more than a few containing a brown, sticky residue. More than likely this liquid leaked into the Metalcraft boxes, leading the company to eliminate using the filled bottles. Most of the "Art Deco" versions we have seen came with bottles that were painted brown on the inside and capped, again manufactured by either Chattanooga Glass or Owens Illinois.

Photograph of a New York window display. For 79¢ the purchaser received six bottles of Coca-Cola, a miniature truck, ten miniature bottles, and a set of nature study cards. The appeal to the youngsters was tremendous. *The Red Barrel*, May 15, 1933.

Manufacturer: Metalcraft | **Model:** Metal wheels | **Composition:** Pressed steel | **Accessories:** Ten 2½" glass bottles

Size: 11" | **Year:** 1931 | **Rarity:** 3 | **Value:** $800.00–1,000.00 | **Reference:** MC1

Close up of White logo on first versions of Metalcraft grille.

All Metalcraft Coca-Cola trucks shared the same A-frame rear bed.

Note: Metalcraft grille variations.

Manufacturer: Metalcraft **Model:** #171 **Composition:** Pressed steel **Accessories:** Ten 2½" glass bottles
Size: 11" **Year:** 1932 **Rarity:** 3 **Value:** $800.00–1,000.00 **Reference:** MC2

Manufacturer: Metalcraft **Model:** #215 working lights **Composition:** Pressed steel **Accessories:** Ten 2½" glass bottles
Size: 11" **Year:** 1933 **Rarity:** 3 **Value:** $1,000.00–1,200.00 **Reference:** MC3

Manufacturer: Metalcraft **Model:** #215 working lights **Reference:** MC3

Manufacturer: Metalcraft | **Model:** "Art Deco" | **Composition:** Pressed steel | **Accessories:** Ten 2½" glass bottles
Size: 11" | **Year:** 1934 | **Rarity:** 5 | **Value:** $1,800.00–2,000.00 | **Reference:** MC4

THE Metalcraft CORPORATION

5101-15 PENROSE AVE
SAINT LOUIS
September 13, 1934

Ridge Coca Cola Bottling Co.,
Winterhaven, Fla.

Gentlemen:

Request after request has been coming in from Coca-Cola Bottling Companies all over for more miniature bottling trucks of the 1933 style.

We have not manufactured that model this year. 1934 calls for something new and different in the truck line -- and we have it.

In the proof attached you have a picture of the new, modern, streamlined chassis, especially designed this year, which we will use for the new Coca-Cola truck. If, in the absence of an illustration of this truck, you will visualize the Coca-Cola body on this new chassis, you have some idea of the smartness of the completed truck. It will be slightly larger than last years and makes a real job.

Of course this change will entail considerably more manufacturing expense, but a comparison of the two trucks certainly justifies the difference in price. This new truck, which we quote to you at 56¢, if merchandised to the department stores, would retail at about $1.50, and even at the 56¢ cost you should be able to merchandise it under $1.00. By reselling to your dealers large quantities of these new trucks, even at a slight profit - the advertising advantages will more than justify the handling.

We would like to group orders from enough plants to enable us to start manufacturing these new, up-to-date trucks at once, so as to be certain of their distribution for the holiday season.

Will you let us have your order for the quantity you desire - subject to approval of sample!

Yours very truly,
THE METALCRAFT CORPORATION
Gene Soden
Sales Manager.

GS:HC

P.S. For your convenience we have enclosed a self addressed card. Kindly use it to send in your order.

NEW YORK EXHIBIT
TWO HUNDRED 5TH. AVENUE

SELL TOYS ALL YEAR FOR GREATER PROFIT

Letter dated September 13, 1934, announcing the new "modern, streamlined chassis" for the "Art Deco" model, which verifies its authenticity.

Price range reflects trucks in 8.5 condition, not MIB. Same truck with original box could substantially increase its value.

SMITH MILLER

SMITTY TOYS

SMITH-MILLER

1944–1945

SMITH-MILLER

In all the great Coca-Cola toy truck collections, you will probably find several Smith-Miller trucks proudly displayed. These beautiful and impressive trucks are undoubtably among the elite of American-made toy trucks. Until now, however, not much has been known about the Smith-Miller Company. In fact, this is one of the most enigmatic toy companies ever; one that has left us with many unanswered questions throughout our many years of collecting. In writing this book we were lucky to have Fred Thompson, current owner of Smith-Miller, share his invaluable knowledge and information with us through several lengthy interviews, enabling us to provide this brief but accurate history of a company that has mystified collectors for years.

The company began in 1944 when Bob Smith, a retired ace with the Flying Tigers during World War II and Matt Miller, who made wooden products, went into partnership to forge Smith-Miller, Incorporated in Hollywood, California. Bob Smith provided more of the financial backing in the venture, while Matt Miller, whose love for trucks, coupled by his woodworking abilities, was the true creative force behind the company. Unlike other toy companies of that era, Smith-Miller, also known as Smitty Toys, only manufactured toy trucks. The logo reads, "Famous Trucks in Miniature." Like the Metalcraft Corporation, most Smith-Miller trucks advertised company names. Among these were Arden Milk, Bekins Van Lines, Kraft Foods, Bank of America, Rexall Drug, Mobilgas and, of course, Coca-Cola.

Again, like Metalcraft, a personal connection prompted Bob Smith and Matt Miller to use Coca-Cola for their first-ever toy truck. According to Fred Thompson, the reason for this was that Bob Smith's wife, known as Babs, had been a Coca-Cola pin-up girl during the early 1940s. This first truck, commonly referred to as the "Wood Smitty," was, as its nickname implies, made completely out of wood, as were the majority of toys produced at that time due to wartime metal shortages and restrictions.

The second Coke truck the company produced, erroneously referred to as the "sandcast," is definitely the victim of a misnomer since the cab was actually aluminum casted in a metal mold instead of sandcasted. The wood Smith-Miller and this successor shared identical rear wooden A-frame beds and displayed the first Smitty Toys logo on their respective cab doors. Between 1946 and 1954 the Smith-Miller company went on to produce three more models of Coca-Cola trucks including the red Chevy in 1946, the red GMC in the late 1940s, and the first yellow, all-metal GMC in 1953. Both the red Chevy and the red GMC had metal cabs and shared identical wood beds displaying Coca-Cola bottles on all four corners. In addition, the GMC version also had "Drink Coca-Cola" decals on the side of the bed, while the Chevy version did not. The Chevy model was the only Coke truck using the second Smitty Toys logo on its cab doors while the GMCs sported the third and final version of the logo. (See photos on page 25–29.) Some collectors believe that there is another version of the GMC Coca-Cola truck with a yellow wood bed and a red metal cab. Smith-Miller did in fact picture this exact truck in their 1947 catalog but, according to Fred Thompson, for reasons unknown, Smith-Miller never produced this truck in an assembly line because none have ever turned up.

By 1950, Bob Smith had sold out his share of Smith-Miller. Soon afterwards, in late 1951, Matt Miller, with a financial-backing partner named Ironson, formed a second company named the Miller Ironson Corporation, known as M.I.C. Miller was hoping to mass produce a more deluxe action truck to compete with companies like Tonka Toys and Buddy "L." Even though Miller left Smith-Miller to focus on M.I.C., the Miller Ironson Corporation only remained in business for about 18 months.

In late 1953, Henry Wolking, who had earlier assumed control of Smith-Miller, acquired M.I.C. His objective was to put both companies under one roof. It was probably Wolking who produced the first yellow all metal Smith-Miller GMC. For the first time ever, these yellow GMCs came equipped with six engraved plastic Coca-Cola crates containing 24 removable Coke bottles, unlike their predecessors which had all come with small wooden blocks, simulating Coke crates. Despite Wolking's arduous work, and much to his dismay, he just could not make the toy business work. He simply closed the doors on this venture and literally let it sit there for the next 25 years. All he ever did was walk through the building daily on the way to his other business (manufacturing pulleys for air conditioners and coolers) which was attached to the rear of the building. Wolking never liquidated Smith-Miller. In fact, through the years, anyone who needed parts for their Smith-Miller trucks could readily find them by simply calling information and checking for the Smith-Miller Company in the Los Angeles white pages. This was exactly what Fred Thompson did when he needed parts for his trucks in 1976. Three years later, Fred Thompson bought both Smith-Miller and M.I.C. from Wolking. According to Thompson, the Smith-Miller warehouse "looked as if everyone had just gone to lunch and never returned." Wolking's other business was still attached to the rear of the Smith-Miller warehouse, which had been off-limits to everyone except Wolking for years.

After buying the company in 1979, Fred Thompson discovered 150 yellow GMC Coca-Cola trucks, which had been painted and decaled in 1953, and were ready for assembly. He assembled them and put them in the original Smith-Miller shipping boxes. The only difference between the trucks sold in 1953, and the ones that were assembled in 1979, were the crates that Thompson used. The original crates had probably run out in the 1950s. The original yellow crates were deeply engraved, injected plastic crates manufactured at the Louis Marx & Company plastic factory. (See photograph on page 29.)Through our research on Marx, we suspect that Marx either destroyed the molds for these crates or turned them over to the Coca-Cola Company at their request. Besides the engraving, another way to distinguish the original crates is by the color of the bottles. The original bottles were light green, almost clear, while the newer bottles are a much deeper emerald green. However, Thompson told us that since he found a supply of the original lighter green bottles in the warehouse, some of the trucks were actually sold with the lighter bottles.

In 1979 Fred Thompson also produced 50 red, all-metal GMC Coke trucks, each one numbered. (See photo on page 30.) These trucks were made entirely from original parts, except for the polished aluminum plate affixed to the rear body panel, which enabled him to use original "New Old Stock" red Coca-Cola decals. The original center panel, made of aluminum as well, was also highly polished in order to affix "N.O.S." Coca-Cola decals. According to Fred Thompson, the reason he made this limited red version was because his son, Tim, then nine, desired a red Coca-Cola truck.

There are still more than 150 unpainted bodies of Coke trucks remaining at the Smith-Miller warehouse today, but Mr. Thompson does not choose to paint or assemble them. As he tells us, he just does not feel right about assembling these trucks . Even though he has all of the materials that he needs, he takes a dim view of capitalizing on something popular just to exploit the market. Unfortunately, collectors need to be aware of less ethical practices, as reproduction Smith-Miller Coca-Cola trucks have surfaced in several areas around the United States. We have observed far too many of these in the course of our travels, including some with unauthorized, reproduced Smith-Miller wooden beds. Fred Thompson agrees with us that it is unconscionable that certain parties are capitalizing on both the names, Smith-Miller and Coca-Cola, by manufacturing these parts that make it possible to turn any Smith-Miller truck into a Coca-Cola truck. This practice is intended solely to mislead the novice collector and, is beginning to greatly compromise the desirability and value of original Smith-Miller Coca-Cola trucks. NOVICE BUYERS BEWARE!

Today, Fred Thompson designs and produces new lines of Smith-Miller trucks but, at this time, usually only 250 to 350 of any particular model, depending on what he feels the market wants. These new Smith-Miller trucks are very collectible in their own right and, like the older trucks, are beautifully designed and of superior quality. In fact, it is not unusual for these trucks to appreciate three to six times their original value in a very short time. To the Coca-Cola truck collector, however, "Nothing Beats the Real Thing." Who knows, maybe Fred Thompson will make a new, limited edition Coca-Cola truck someday! For more information on Smith-Miller, look for Fred Thompson's upcoming book on his company sometime in the near future.

Manufacturer: Smith-Miller **Model:** All wood **Composition:** Wood **Accessories:** 16 wooden blocks stamped Coca-Cola
Size: 14" **Year:** 1944 **Rarity:** 5 **Value:** $2,500.00–2,800.00 **Reference:** SM1

Rear and side view of Smith-Miller Coca-Cola truck from 1944. **Reference:** SM1

Actual molds used to produce 1945 Ford, often erroneously refered to as "sand cast." *Photo courtesy Fred Thompson.*

Manufacturer: Smith-Miller **Model:** Ford A-frame **Composition:** Wood and aluminum **Accessories:** 16 wooden blocks stamped Coca-Cola
Size: 14" **Year:** 1944–45 **Rarity:** 4 **Value:** $1,800.00–2,000.00 **Reference:** SM2

Reference: SM2

Reference: SM2

Reference: SM2

Manufacturer: Smith-Miller
Model: Chevrolet
Composition: Cast metal, wood
Accessories: 16 wooden blocks stamped Coca-Cola
Size: 14"
Year: 1945–46
Rarity: 3
Value: $1,200.00–1,400.00
Reference: SM3

Reference: SM3

Reference: SM3

Reference: SM3

Manufacturer: Smith-Miller
Model: GMC
Composition: Cast metal, wood
Accessories: 16 wooden blocks not always stamped Coca-Cola
Size: 14"
Year: 1947–53
Rarity: 3
Value: $1,100.00–1,300.00
Reference: SM4

Reference: SM4

Reference: SM4

Manufacturer: Smith-Miller **Model:** GMC yellow **Composition:** Metal
Accessories: 6 Coca-Cola etched plastic cases w/24 light green bottles
Size: 14" **Year:** 1953–54 **Rarity:** 4
Value: $850.00–950.00 **Reference:** SM5

Price range reflects trucks in 8.5 condition, not MIB. Same truck with original box could substantially increase its value.

Manufacturer: Smith-Miller **Model:** GMC yellow **Composition:** Metal **Accessories:** 6 Coca-Cola stenciled plastic cases w/24 bottles
Size: 14" **Year:** 1953–54, assembled in 1979 **Rarity:** 3 **Value:** $650.00–850.00
Reference: SM6 **Note:** Bottles are light green but more often emerald green

Reference: SM6

Unpainted GMC beds manufactured in 1954 are still in Smith-Miller warehouse today.

Note: Deeply etched case on left, c.1954; stenciled case on right, c.1979. (See page 21.)

Manufacturer: Smith-Miller **Model:** GMC (50 produced) **Composition:** Cast metal
Accessories: 6 red plastic cases; 24 emerald green bottles **Size:** 14"
Year: 1979 **Rarity:** 4 **Value:** $1,200.00–1,500.00
Reference: SM7 **Note:** Cases have new Coca-Cola logo

All 50 red GMC trucks produced in 1979 were serial numbered #1–50 on chassis.

The highly polished aluminum rear panel was the only part manufactured in 1979. It was made to affix the new old stock decals. See page 21.

BUDDY "L"

1948–1969

BUDDY "L"

Buddy "L" toy trucks are among the best known and certainly the oldest trucks in the American toy industry. The force behind Buddy "L" was Fred A. Lundahl who formed the Moline Pressed Steel Company in East Moline, Illinois, in 1910. The Moline Pressed Steel Company manufactured parts for truck and automobile makers like International Harvester, as well as, parts for the booming farm machinery industry of the time. Economic prosperity was short-lived as contracts dwindled. Less than a decade later Moline Pressed Steel faced severe financial problems. These problems continued to plague the company (later known as the Buddy "L" Corporation) for most of its 70 years in business.

Desperately in need of capital, Fred Lundahl approached International Harvester in the late 1910s with the concept of making miniature models of their trucks as promotional items to stimulate sales. Lundahl had been making toys for his son out of scrap metal for several years before the economic slump hit, which probably gave him this idea. International Harvester rejected his proposal, but Lundahl, unaffected, decided to pursue his dream of making toys and in 1923, Moline Pressed Steel went into the toy business exclusively.

Although the company's official name remained Moline Pressed Steel, Lundahl named his special line of toys after his son, Arthur Brown "Buddy" Lundahl. Hence the name Buddy "L" was born and still remains. Lundahl wanted durable, high quality toys for his son, as well as all children. His products appealed to parents all over the United States who were fed up with the overall quality of toys available on the market. According to *The New Book of Buddy "L" Toys*, Buddy "L" toys were an instant success with the "first heavy, indestructible pressed steel toys and the first to combine highest quality, realistic operation, amazing durability and innovative design in playthings." Lundahl, who often referred to himself as the "daddy" of Buddy "L" in company advertising, liked to demonstrate the durablity of his toys by standing on toy trucks to prove that they could withstand his 220 lb. weight. Compared to the inexpensive tin trucks on the market, Buddy "L" trucks were somewhat more expensive, but Lundahl made sure prices remained competitive.

Throughout the 1920s, Lundahl continued to expand his line of steel toys, and while many competitors surfaced, the quality of the Buddy "L" line remained unparalleled. At the end of the decade, a serious blow was dealt to Moline Pressed Steel with the advent of the Great Depression and, subsequently, with the death of Fred Lundahl in 1930. The same year, Moline Pressed Steel was officially renamed the Buddy "L" Manufacturing Company. This was the first of many name changes that would continue through the years, including the East Moline Toy Company, the Buddy "L" Company, and the Buddy "L" Toys Division of the Moline Pressed Steel Company, until the company finally settled on the Buddy L Corporation in 1960 (it was then that the quotation marks around the "L" were officially dropped.)

Henry Katz brought new life to Buddy "L" when he bought the company in 1941. Katz was the most influential person in the company's history aside from Fred Lundahl. Since 1933, he had been national sales representative for Buddy "L" and was a master salesman. He directed the company through still another rough period during World War II when steel shortages were at an all time high. Katz purchased a woodworking plant in Glens Falls, New York, where wood trucks were produced for the first time in the company's history. It was at this plant that Buddy "L" introduced its first Coca-Cola truck, often referred to simply as the Wood Buddy "L," in the late 1940s. The East Moline plant was contracted by the government during the war to produce only military items between 1942 and 1945. The Buddy "L" November 1945 catalog proudly announced, "Our war work is finished, and we are again at the business of making our famous line of steel playthings." The Glens Falls factory continued making wooden toys until 1949, but consumers felt they were no match for their steel counterparts. Nevertheless, Buddy "L" continued to promote its two distinct divisions as Buddy "L" Wood Toys; Glens Falls, New York, and Buddy "L" Steel Toys; East Moline, Illinois, for several years.

An already weak and disgruntled toy industry soon realized that steel would again become scarce due to the advent of the Korean War. Again, Buddy "L" searched for an alternative and began manufacturing a few plastic toys in 1952. By the late 1950s, International Harvester was no longer being used and the GMC body style was introduced. Beginning in the early 1960s, the styling of both International Harvester and GMC was abandoned for a Ford model, especially popular at the time. One can see this progression of body styles in the Coca-Cola toy bottling trucks. The first metal Coke truck is known as the 3-tier (model 5536), making its first appearance in 1955 with International Harvester styling. This same International Harvester styling was again used in 1957, yet this was a newly restyled version (model 5546) with two tiers instead of three and with two grille variations. In

Manufacturer: Buddy "L" **Model:** Wood **Composition:** Wood **Accessories:** 8 wooden blocks **Size:** 19" **Year:** 1947–48
Rarity: 5 **Value:** $2,500.00–3,000.00 **Reference:** BL1

Front view of Wood Buddy "L."
Reference: BL1

1958, Buddy "L" opted for the GMC 550 body style (model 5646). It is interesting to note that this GMC Coca-Cola truck, like its predecessors, was painted yellow, yet the following year this truck appeared in an interesting orange version unique to this particular edition. There has often been speculation that this orange version was the result of a factory error in the paint, but it has been conclusively proven in the Buddy "L" archives that this truck was painted as planned from the beginning – why orange remains a mystery!

The 1960s saw the official change of the Buddy "L" logo from the round logo to a new slanted rectangular-shaped logo. This new logo changed very slightly over the years. The word Buddy "L" always appeared in white italics on a red background either framed in black before 1963 or framed in white after 1963. While examining the trucks closely, it is noticeable that there are slight variations in the size of the logo. (See photographs this section.) Additionally, although unsuccessful in obtaining the license to use the Ford trademark on their new trucks, the 1960s Coca-Cola route trucks were in fact designed with Ford body styling (the now famous model 5426). These Ford look-alikes were produced for a decade with very slight modifications, most noticeable were minor features like grilles, side mirrors, and bumpers. The first model 5426, for example, had a yellow wrap-around bumper in 1960, while the 1961 model was equipped with a white plastic grille guard, often advertised as a model offering "protection for parents' furniture."

Buddy "L" started manufacturing a less expensive polysteel truck in the late 1950s. The first polysteel Coke truck (model 5216) appears in 1961. Contrary to what the name implies, the trucks were constructed completely of polyethylene plastic except for a steel plate mount holding the cab and body together. The second and only other polysteel Coke truck, introduced in 1962, differed only slightly from the 1961 model.This newer model lacked cab door decals and a "Come-Back Motor," which allowed the truck to jut forward when pulled backwards. In addition to polysteel, Buddy "L" also experimented with spring suspension, to make Buddy "L" trucks "just like the real trucks," from 1961 to 1964. In 1961 and 1964, the spring suspension was located only in the front axles of the model 5426 Coca-Cola route truck while, from 1962–1963, the model 5426 truck had spring suspension in both front and rear axles.

By this time, competition among toymakers was fierce. The Buddy "L" Corporation no longer manufactured its high quality line of solid steel trucks as it had in its golden years of the 1920s and 1930s. Henry Katz still strived to attain the highest quality possible; however, different materials had to be introduced so that these trucks could remain competitively priced. Foreign production was also necessary to stay in the toy truck market. In 1966 the first Buddy "L" import arrived from Japan. Within nine years, all Buddy "L" toys were destined to be manufactured in Asian countries. From 1975 onward, toy trucks were made to Buddy "L" specifications principally in Japan.

Booklet page featuring barbecue grills manufactured by Buddy "L" in 1969.

In 1969, Buddy "L" moved its headquarters from East Moline, Illinois, to Neosho, Missouri, where the company entered the barbecue grill business. It was believed that with this move, the plant could be earning money year-round since both the barbecue grill and toy industries were seasonal. In 1973, Henry Katz sold the Missouri factory to the Sunbeam Company, which continued to manufacture barbecue grills. As far as the toys were concerned, Katz maintained the research and development portion of the Buddy "L" Corporation and set up its headquarters in New York City. However, he contracted out the manufacturing of the toys. Sadly, this proved no more successful, and the corporation filed for bankruptcy in 1985. In 1990, St. Lawrence Manufacturing (SLM, Inc.), based out of Montreal, Canada, bought Buddy "L" along with Coleco and established a base of operations in Gloversville, New York. This is probably the reason for the similarity between Coleco and Buddy "L" Coke trucks.

For more information on post-1969 Buddy "L" trucks see Modern Era/New Logo chapter.

Progression of logos used on Buddy "L" trucks from 1955–1969.

1955–1958

1958–1960

1960–1963

1963–1965

1965–1969

Manufacturer: Buddy "L" **Model:** #5536 International 3-tier **Composition:** Pressed steel
Accessories: Metal hand truck, 8 yellow/red cases **Size:** 14¼" **Year:** 1955 **Rarity:** 2 **Value:** $250.00–350.00
Reference: BL2

Manufacturer: Buddy "L" **Model:** #5546 International 2-tier **Composition:** Pressed steel
Accessories: 2 metal hand trucks, 8 yellow/red cases **Size:** 14¼" **Year:** 1957 **Rarity:** 2
Value: $250.00–350.00 **Reference:** BL3

Manufacturer: Buddy "L"
Model: #5546 International
Composition: Pressed steel
Accessories: 2 metal hand trucks, 8 yellow/red cases
Size: 14¼"
Year: 1958
Rarity: 2
Value: $250.00–350.00
Reference: BL4

Manufacturer: Buddy "L"
Model: #5546 International
Composition: Pressed steel
Accessories: 2 metal hand trucks, 8 yellow/red cases
Size: 14¼"
Year: 1959
Rarity: 2
Value: $250.00–350.00
Reference: BL5

Price range reflects trucks in 8.5 condition, not MIB. Same truck with original box could substantially increase its value.

Note: Cab shared by both Model #5536 and #5546.

Note: Rear view of Model #5536.

Manufacturer: Buddy "L"
Model: #5646 GMC
Composition: Pressed steel
Accessories: 2 metal hand trucks, 8 green/red cases, conveyor belt
Size: 14¼"
Year: 1957–58
Rarity: 3
Value: $400.00–500.00
Reference: BL6

Decal on under panel of GMC models.

Manufacturer: Buddy "L"
Model: #5646 GMC - orange
Composition: Pressed steel
Accessories: Same as BL6
Size: 14¼"
Year: 1958
Rarity: 4
Value: $650.00–750.00
Reference: BL7

Manufacturer: Buddy "L"
Model: #5426 w/yellow wrap around bumper
Composition: Pressed steel
Accessories: Same as BL6, plus a black left side mirror
Size: 15"
Year: 1960
Rarity: 4
Value: $350.00–450.00
Reference: BL8
Note: This was the first version of the new "Ford style" and it is the rarest.

Manufacturer: Buddy "L"
Size: 12¼"
Reference: BL9
Model: #5216 polysteel
Year: 1961
Note: Has "Come-Back Motor"
Composition: Molded plastic
Rarity: 5
Accessories: Metal hand trucks, 8 green/red cases
Value: $400.00–600.00

Manufacturer: Buddy "L"
Model: #5216 polysteel
Composition: Molded plastic
Accessories: 8 green/red cases
Size: 12¼"
Year: 1962
Rarity: 5
Value: $400.00–600.00
Reference: BL10

Manufacturer: Buddy "L"
Model: #5426 spring front w/plastic grille
Composition: Pressed steel
Accessories: 2 metal hand trucks; 8 green/red cases
Size: 15"
Year: 1961
Rarity: 2
Value: $175.00–275.00
Reference: BL11

Manufacturer: Buddy "L"
Model: #5426 single axle spring
Composition: Pressed steel
Accessories: 2 metal hand trucks, 8 green/ red cases
Size: 15"
Year: 1962-64
Rarity: 2
Value: $175.00–275.00
Reference: BL12

Buddy "L" was obviously trying to appeal to Moms and Dads with their furniture safe plastic bumper.

Wish Book or mini-catalog featuring #5426. It was enclosed in toy boxes or found at neighborhood toy store counters, c.1965–1969, size 5½" x 3½".

Manufacturer: Buddy "L"
Model: #5426 chrome grille
Composition: Pressed steel
Accessories: 2 metal hand trucks, 8 green/red cases
Size: 15"
Year: 1965-69
Rarity: 2
Value: $175.00–275.00
Reference: BL13

Coca-Cola
MADE IN UNITED STATES OF AMERICA
MAR
TOYS
MARX
1947–1967

MARX

Marx trucks offer collectors the most variety of any of the American toy manufacturers. As with all the preceding companies, Louis Marx & Company was the product of one man's great genius. Louis Marx, who formed the company in 1919 at age 23, started out in the toy business in his mid-teens. Marx began his career working for Ferdinand Strauss, becoming director of the Strauss Toy Company at age 20. Louis Marx, above all a brilliant businessman, turned everything he touched into gold. It soon became apparent, however, that Marx and Strauss envisioned very different things for the company. It was these irreconcilable differences that gave Marx incentive to start his own business with his brother, David, who became the public relations man in the new venture. By age 26, Louis Marx was already a millionaire and recognized the world over as the "Toy King," a title previously owned by Strauss.

Marx started his company in 1919, but did not actually start manufacturing toys until he bought his first plant in 1921. During those first two years, Louis and David Marx acted solely as middlemen between toy manufacturers and wholesale buyers. This provided Marx with a great opportunity to explore the market and create a niche for himself. He could not have entered the toy business at a better time. Before World War I, Germany had a virtual monopoly on the toy industry. In fact, more than 85% of all toys bought by Americans in the early 1900s came from either Germany, France, Austria, or England. By 1925, however, the United States toy industry grew some 900% and became the largest in the world due in no small part to companies like Louis Marx and Buddy "L." It is also interesting to note that Marx bought many of Strauss' toy molds after World War I when Strauss went out of business.

The Louis Marx Company became a favorite with toy buyers because of Marx's commitment to providing high quality toys at the lowest possible prices. In order to keep prices down, Marx established a high volume sales policy. The only drawback to this was that smaller businesses like local five and dime stores could not afford to buy such large quantities. Often these smaller retailers would form cooperatives in order to buy Marx products, which they then divided amongst themselves.

At first, Marx only produced metal toys, and most collectors agree that Marx produced some of the nicest lithographed steel and aluminum trucks ever. In fact, Marx was the first toy company in the United States to use the lithographic process on Coca-Cola toy trucks. Louis Marx, however, was no more immune to the effects of the metal shortages during World War II than was the rest of the toy industry. Government Office of Production Management restrictions prevented any toy from containing more than 10% metal. Marx manufactured a few wood toys at this time, as well as cardboard and paper playthings (none of which were Coca-Cola). During the height of the United States' involvement in the war, Marx actually ceased toy production for four years in order to fulfill government military contracts, mainly producing parts for bombs. In fact, a huge four to six foot long bombshell was even mounted on a wall in one of the Marx plants in Pennsylvania as a gruesome memorial to the war.

By 1948, Marx had already started experimenting with plastics. He was not the pioneer in plastics he had been with tin lithography, but soon recognized there was a new trend in the market. The first plastic Coca-Cola truck was advertised in Marx's 1950 Christmas catalog. Marx's involvement and quick mastering of plastics exemplifies the philosophy that made him successful. He always believed that technology was the key to success since better technology resulted in higher quality, greater efficiency, and ultimately a better product. By 1952, the use of plastic was in full swing at Marx factories as Marx had now become a leader in plastics technology.

By 1950, the Marx Company had become the largest toy manufacturer in the world with approximately 15% of the entire wholesale toy market. By the middle of the decade, Marx oversaw six different factories in the United States – the three most famous located in Erie and Girard, Pennsylvania, and Glendale, West Virginia. These three plants are probably the most often referred to by collectors. The Girard plant opened in the mid-1930s and was best known for its toy train production. The Glendale plant, a former aircraft factory, manufactured the more primitive heavy-gauge metal toys, and the Erie plant manufactured metal toys from the mid-1920s to the late 1940s when production of plastics was added as well. Erie was probably Marx's most important plant as it also served as headquarters for Marx's research and development department until its closing in 1975.

In addition to these three main American plants, Marx established several plants overseas including Canada, Mexico, Brazil, France, Germany, Wales, Hong Kong, Japan, South Africa, and Australia. To Coca-Cola truck collectors, the most important foreign plants were located in Canada, Mexico, and Japan. Marx had different arrangements with each plant. For instance, the Canadian plant did basic assembly and paint-type work with parts manufactured in Erie and Glendale. In contrast, the Japanese plant, operating under the name Linemar, produced its own toys. Most of the Japanese toys were mechanical, wind-up, or friction, and Linemar produced several very exciting Coca-Cola trucks. (See North America and Japan chapters for more information.)

Believing that toys were a "young man's business," Louis Marx sold his company to the Quaker Oats Company in 1972. Once the "King" stepped down, however, the kingdom collapsed. The company simply could not survive without Marx and his "midas touch." Quaker Oats continued losing money for the next four years, finally selling Marx to one of Europe's largest toy manufacturers, Dunbee-Combex, Ltd. Unfortunately, Dunbee-Combex-Marx, as it had been renamed, fared no better and went into bankruptcy in 1980. Louis Marx lived long enough to see his former company dissolved. He died in 1982.

Louis Marx & Company produced nine distinct models of Coca-Cola trucks during the 20-year span between 1947 to 1967. The first truck produced was the Sprite Boy Stake Truck, model #991, introduced in the

Manufacturer: Marx **Rarity:** 3 **Model:** #991 stake **Value:** $450.00–650.00 **Composition:** Pressed steel **Reference:** MX1 **Size:** 20¼" **Year:** 1948–49

Price range reflects trucks in 8.5 condition, not MIB. Same truck with original box could substantially increase its value.

Note: Rear decal of first version #991.

late 1940s and continuing into the early 1950s. Concurrently produced, were several models made of plastic. Eventually, the heavy-stamped steel stake trucks were phased out, as they were no longer competitive with materials that followed. This brought about the introduction of the tin lithographed models with or without added decals beginning with the famous mid-1950s model, easily recognized by a large #21 located over the front wheel wells. These #21s, as they are called, were abundant throughout the mid-1950s but later gave way to a new generation of all-lithographed stake trucks including the #1088 and the #1089. The last in the line was the model #1090, Marx's triumphant masterpiece of toy design. This magnificent toy sported lithographed cases as well as 16, 18, or 32 plastic cases with 24 miniature green bottles per case. These beautiful and colorful toy trucks are among the most collectible and valued by collectors today. We are proud to feature the most complete group of Marx Coca-Cola toy trucks ever assembled in any book.

Cover and page of the 1953 Christmas catalog from Aldens Department Store, Chicago. Note that the Marx #991 was available in assorted colors. See description below.

Truck description from the above Aldens catalog:
"9. COCA-COLA TRUCK. Heavy gauge steel, stake design. Shiny radiator grille, bumper. Dummy headlights. Baked-on enamel finish in assorted bright colors. Authentic Coca-Cola decals on both sides of truck body. Metal lithographed wheels. 20½ x 6⅛ x 6¼ - in.
352 Y 5401 Shipping weight 6 lb. 4 oz. .... 1.92"

Manufacturer: Marx
Model: #991 stake, no rear decal
Composition: Pressed steel
Size: 20¼"
Year: Early 1950s
Rarity: 2
Value: $350.00–550.00
Reference: MX2

Manufacturer: Marx
Model: #991
Composition: Pressed steel
Size: 20¼"
Year: Early 1950s
Rarity: 2
Value: $450.00–650.00
Reference: MX3

Manufacturer: Marx
Model: #991
Composition: Pressed steel
Size: 20¼"
Year: Early 1950s
Rarity: 4
Value: $550.00–750.00
Reference: MX4

Manufacturer: Marx **Year:** 1954–56
Model: #21 **Rarity:** 2
Composition: Tin **Value:** $350.00–450.00
Accessories: 6 plastic cases **Reference:** MX5
Size: 12½"

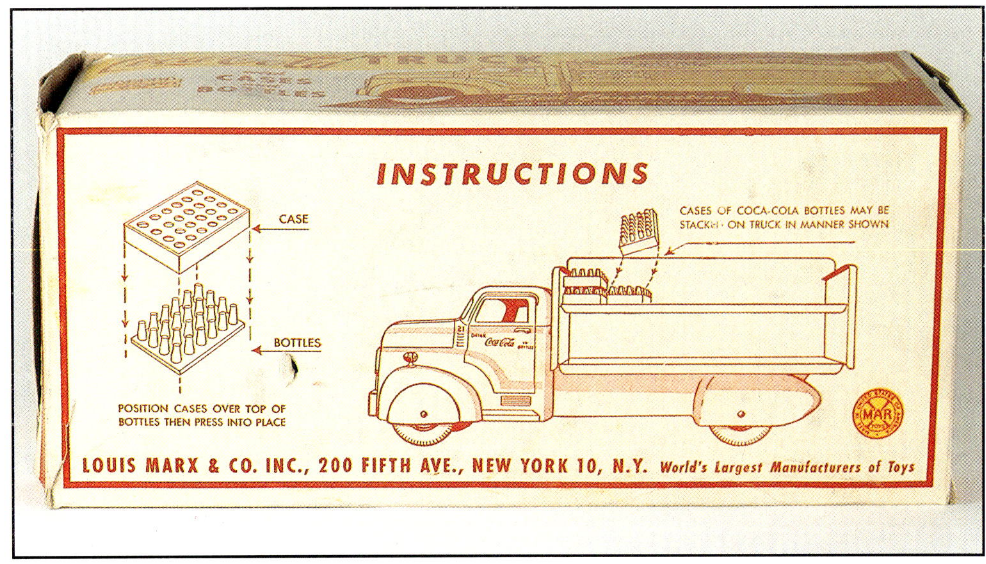

Bottom of box showing assembly of crates for MX5.

Manufacturer: Marx **Model:** #21 **Composition:** Tin **Accessories:** 6 plastic cases, yellow or red metal hand truck
Size: 12½" **Year:** 1954–56 **Rarity:** 2 **Value:** $350.00–450.00 **Reference:** MX6

Manufacturer: Marx **Model:** #21 w/horns **Composition:** Tin **Accessories:** 6 plastic cases, yellow or red metal hand truck
Size: 12½" **Year:** 1954–56 **Rarity:** 3 **Value:** $375.00–475.00 **Reference:** MX7

Manufacturer: Marx **Model:** #1088 **Composition:** Tin **Size:** 18¾" **Year:** 1956
Rarity: 3 **Value:** $450.00–650.00 **Reference:** MX8

LOUIS MARX & CO.
200 Fifth Ave., New York

GLENDALE DIVISION
12/20/56

NO. 1088 COCA COLA TRUCK

Sizes, 18-3/4" long x 6-3/8" x 5-1/2" high.

FACTORY SPECIFICATIONS:

Model No. 2472-5B
Utilizing the #3A cab
PL-834 Polyethylene wheels
1 - 1002 stake body with Coca Cola litho
4 - tinplate wheel discs
1set - Air horns - ?

Glendale factory spec sheet describing model #1088, c. 1956.

Manufacturer: Marx
Year: 1956–57
Model: #1089
Rarity: 3
Composition: Tin
Value: $425.00–625.00
Size: 17½"
Reference: MX9

Actual Marx #1089 pictured in *Goldstein's Coca-Cola Collectibles*. The left rear is on a block of wood. This truck was purchased mint in box in the late 1960s, but the left rear wheel was missing and a block of wood placed there by Shelley Goldstein has remained ever since.

Manufacturer: Marx **Model:** #1090 **Composition:** Tin **Accessories:** 15, 18, or 24 cases (depending on year), hand truck
Size: 17½" **Year:** 1956–57 **Rarity:** 3 **Value:** $750.00–1,000.00 **Reference:** MX10

Price range reflects trucks in 8.5 condition, not MIB. Same truck with original box could substantially increase its value.

LOUIS MARX & CO.
200 Fifth Ave., New Yo

GLENDALE DIVISION
3/14/57

NO. 1090 COCA COLA TRUCK

Measures 17-1/2" long x 5-3/4" wide x 6-3/4" high.

Modern replica of Coca Cola with regulation colors and authentic markings.

Truck of sturdy all-steel construction cab-over-engine design cab with radiator, grille, dummy headlights and sturdy xxx front bumper.

The body portion measuring 11-1/2" long contains 15 miniature Coca Cola cases, each case containing 24 xxxxxxxx miniature plastic bottles. Miniature hand truck is also supplied for the delivery of the cases.

The lower portion of the body is lithographed steel insert simulating rows and rows of Coca Cola xxxxxxxxxxxxxxxx bottles and has a top upright panel running through the center of the truck with appropriate trade mark and design.

A full length heavy-gauge reinforced chassis. Truck equipped with four heavy-duty rubberized tires with shiny hub discs.

PACKED: xxxxxxxxxxx Each in a box,
1/4 dozen to shipping carton,
Weight per carton - 18-1/4 lbs.

FACTORY SPECIFICATIONS:

Model No. 2472-5A

No. 3B Cab in regulation Coca Cola litho (closed fenders)
1 - Long chassis - yellow
1 - Basic Coca Cola body of #1090 - yellow w/ Coca Cola insert
Coca Cola decal on rear of body
15 - PL-31 and 32 Coca Cola cases & bottles, case yellow, stamping red
1 - PL- Plastic hand truck — red
4 - PL-834D Wheels with shiny hub caps

NOTE: No Air Horns

This is the basic 1090 of last year with revisions leaving off the Air Horns.
Change the polyethylene wheels and reduce cases from 18 to 15.
Change to 3B cab

Copy of factory spec sheet from Glendale plant describing #1090, c. 1957. Please note first issue of Marx #1090 had lithographed horns on the cab, while subsequent issues did not.

Marx Original

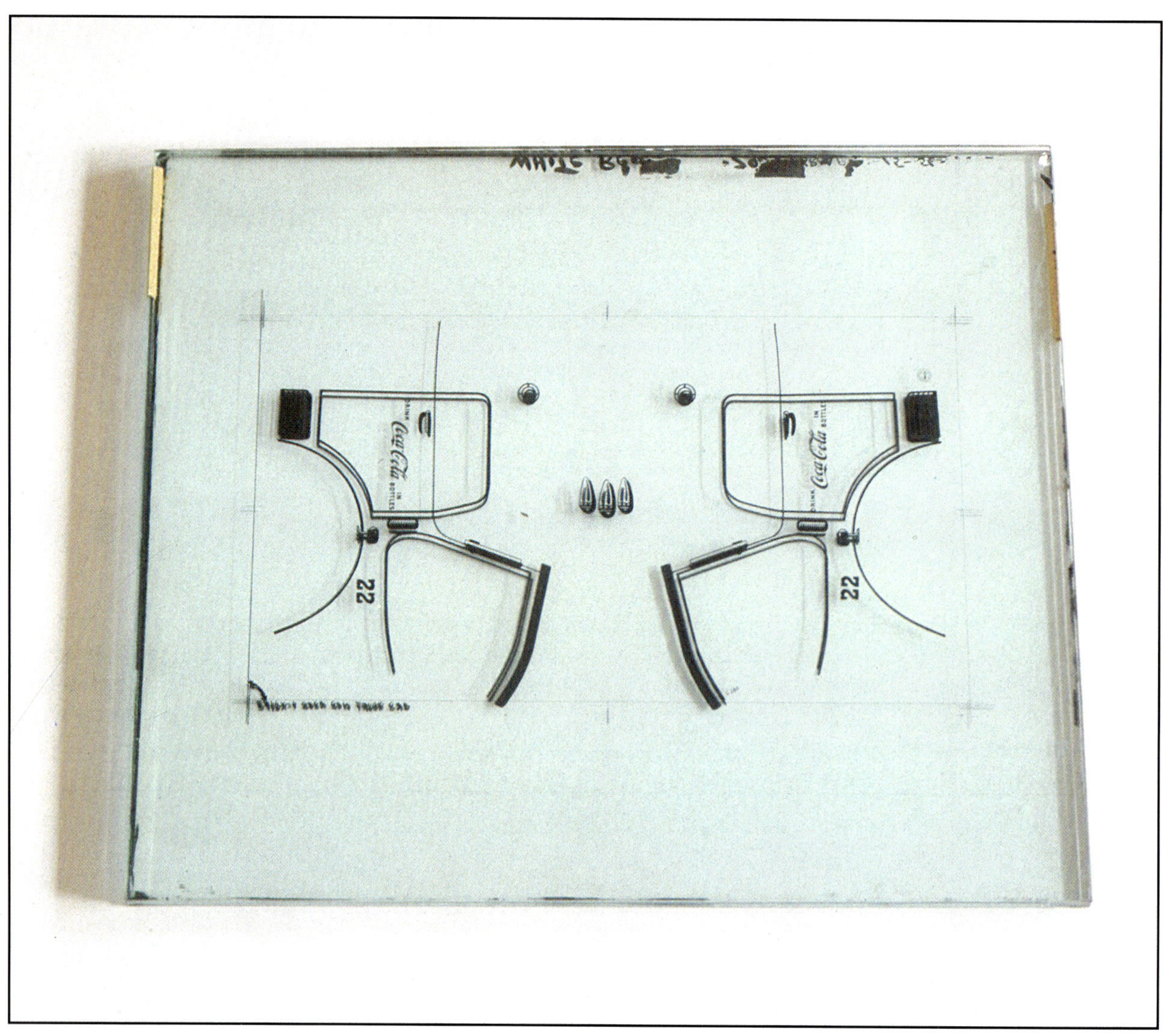

On these pages, photos show one of a kind glass plates used by Marx to lithograph the #1090 Coca-Cola truck. They were purchased at the Glendale, West Virginia, plant along with boxes of parts in the late 1970s. These photos show four superimposed glass plates; each plate used for the screening of a different color of ink. It is a rare opportunity to see the actual plates that were used to produce the truck you may have in your collection. We wish to thank our friend, John Varga, for lending these irreplaceable and historically important pieces for us to photograph.

Lithography Plates

Lithography plates (4) used for the Marx #1090 Coca-Cola truck shown below. **Reference:** MX10

Reference: MX10

Plastics

Although the advent of injected plastics in the late 1940s revolutionized the toy industry, it also seriously jeopardized the metal toy industry. In the early 1950s, toy manufacturers who wanted to remain competitive were forced to explore the possibilities of manufacturing plastic toys, as these could be mass produced at a fraction of the cost since they required fewer production tools and far less labor. As mentioned before, Louis Marx was not a pioneer in the plastic toy industry; however, he was instrumental in the advancement of plastics technology once he started up with plastic toy production.

Early on, the plastics process had not yet been perfected, and Marx had a difficult time with the first plastic toys he produced. Among these problems was the difficulty in releasing the plastic from the molds, and also preventing the warping of the long, straight pieces that made up the bigger toys. These initial problems caused the first plastic toys to be marginal at best. Additionally, finding the right softening components, as well as the ideal temperature for cooling the plastics, made it almost impossible to mass produce the trucks without substantial losses. It is hard for collectors today to find the earlier, brittle plastic trucks that aren't broken or cracked.

Newspaper sales ad for Macy's, c.1950. Note the absence of doors, irrefutably proving that this truck did in fact come with or without doors.

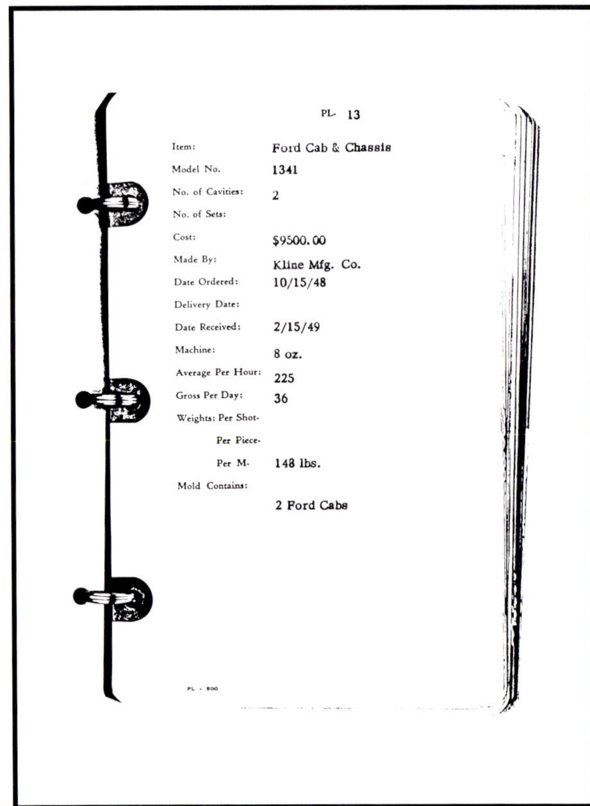

PL- 13

Item:	Ford Cab & Chassis
Model No.	1341
No. of Cavities:	2
No. of Sets:	
Cost:	$9500.00
Made By:	Kline Mfg. Co.
Date Ordered:	10/15/48
Delivery Date:	
Date Received:	2/15/49
Machine:	8 oz.
Average Per Hour:	225
Gross Per Day:	36
Weights: Per Shot-	
Per Piece-	
Per M-	148 lbs.
Mold Contains:	2 Ford Cabs

Typical prospectus document from Marx's research & development department.

By the mid to late 1950s, however, Marx's research and development department's commitment to continually improving plastics technology resulted in the adoption of several new methods for making durable plastic toys. Among these advances were blow molding and sonic welding – two processes pioneered by Louis Marx & Company. Blow molding, in particular, was responsible for revolutionizing plastic toy manufacturing. This process involved the use of molds and air drills to blow molten plastic into a desired shape, which basically made the making of toy trucks a one-step process.

Earlier problems with plastics could explain why some of the first plastic trucks had doors while others did not. We just are not sure; however, this could certainly explain why the mysterious prototype on page 62 may have never been produced. Because of its size, the straight plastic pieces required to assemble the bed may have been too long to avoid warpage. Louis Marx & Company records, however, indicate that up to 180,000 of these trucks were actually ordered to be manufactured. This is a big mystery as far as truck collectors are concerned, since not one of these has ever surfaced. Companies often make prototypes that are never manufactured for one reason or another. This is what probably happened in the case of this mystery truck. It seems unlikely that if 180,000 were produced, none survived.

Also interesting to note is that many molds were eventually sent to Marx's Mexican plant in later years. This explains why the plastic Mexican Marx trucks made from these molds and sold under the name Plasti-Marx, were still being made well into the 1960s. In fact, some of these plastic toy trucks have surfaced in even more exotic places like South Africa, where they were also produced in the late 1960s.

We have recently seen the truck pictured on page 58 (Reference MX16) appear in large quantities. Made from the original molds, it was intended to be a *MARXI* COLA truck. We have further observed that these trucks have been sold to unsuspecting collectors, under the pretense that they were recently discovered Coca-Cola new-old stock. This is definitly not the case, and these trucks are being misrepresented solely to deceive people and to capitalize on Coca-Cola's popularity. Here are a few ways to distinguish these reproductions from the originals: 1) Originals are made from a much brighter, shinier, and more brittle yellow plastic and were *never* produced in red. The new one however is duller, softer, and is available in yellow and red. These reproductions are often in a box and come with Coca-Cola decals. Please note that the original had no decals. It usually, but not *always*, had a cardboard insert. If you see one of these trucks for sale and are unsure, remember – *Caveat Emptor* – Buyer Beware!

Manufacturer: Marx
Model: Ford style w/doors
Composition: Plastic
Accessories: 6 plastic cases, sometimes a hand truck
Size: 11"
Year: 1950–54
Rarity: 3
Value: $225.00–325.00
Reference: MX11

Manufacturer: Marx
Model: Ford style (rare red wheels)
Composition: Plastic
Accessories: 6 plastic cases, sometimes a hand truck
Size: 11"
Year: 1950–54
Rarity: 5
Value: $350.00–450.00
Reference: MX12

Manufacturer: Marx
Model: Ford style w/engraved logo
Composition: Plastic
Accessories: 6 plastic cases, sometimes a hand truck
Size: 11"
Year: 1950–54
Rarity: 4
Value: $400.00–500.00
Reference: MX13

Toy truck advertisement from Spiegel Catalog, Christmas edition, c. 1950.

"**Coca-Cola Set**
Fun galore for make-believe truck drivers! Look at all you get: large yellow plastic truck. . .6 miniature cases of "Coke"–filled with imitation, non-removable bottles. PLUS 2-wheel hand-truck to put in big truck while travelling!
35 J 5102. (1 lb. 8 oz.) .... **1.39**"

Rear view showing engraved logo. This model sometimes came with engraved doors.
Reference: MX13

Manufacturer: Marx **Model:** Chevy style **Composition:** Plastic **Accessories:** 6 cases, sometimes a hand truck
Size: 11" **Year:** 1950-54 **Rarity:** 2 **Value:** $225.00–325.00 **Reference:** MX14

Manufacturer: Marx **Model:** "Snub nose" **Composition:** Plastic **Accessories:** 6 plastic cases, sometimes a hand truck
Size: 11" **Year:** 1950-54 **Rarity:** 4 **Value:** $500.00–600.00 **Reference:** MX15

Manufacturer: Marx **Model:** Cardboard inserts **Composition:** Plastic **Accessories:** 6 plastic cases, lithographed cardboard inserts
Size: 10" **Year:** 1956 **Rarity:** 3 **Value:** $350.00–450.00 **Reference:** MX16

Manufacturer: Marx **Model:** Same as MX16 but without cardboard inserts **Composition:** Plastic
Accessories: 6 plastic cases **Size:** 10" **Year:** Mid-1950s **Rarity:** 3 **Value:** $225.00–325.00
Reference: MX16a **Note:** See reproduction warning on page 55.

Prototypes

Like other companies, Marx produced many prototype Coca-Cola trucks that, for one reason or another, were never produced. Most of the prototypes that we have seen were made in the 1950s and 1960s and surfaced in the early 1970s when Louis Marx sold his company. Prototypes are of special interest, as they offer an insight into the creative minds of the toymakers. Why certain trucks were not produced is often unclear. Perhaps the reason why the 1967 version of the #1090, Model #1128 (see page 61), was never produced was that Marx was unable to get the license. The Coca-Cola Company was undergoing massive advertising changes in preparation for the introduction of their new modern logo. Otherwise, we are not sure why some of these other trucks were not produced. Interestingly though, the plastic prototype seen on page 60 was never produced as a Coke truck but, rather, as a famous Pepsi-Cola truck instead.

As Marx prototypes surfaced on the market, they often caused speculation and rumors as to their appearance, and many stories circulated about them. One of the more interesting ones that we have heard through the years is that the 1967 Gray Cab (#1089) prototype on page 60, had actually been produced as a replica of a vehicle used to haul sugarcane on Coca-Cola sugar plantations. Although this was believed to be true, in actuality it was merely a prototype. It is probable that several prototypes were made for each model, explaining why we have sometimes seen more than one.

It is very hard to assign a value to these scarce toys since there is no comparison. Therefore, we use a not applicable (N/A) as the value for some of these prototypes. One thing is for certain – they are extremely rare and valuable, and are a great addition to any collection.

Note: Typical markings found on Marx prototypes; handpainted "X"perimental ID code with production dates.

Manufacturer: Marx **Model:** #3820X **Composition:** Plastic **Size:** 7½"
Year: 1950–54 **Rarity:** 5 **Value:** N/A **Reference:** MX17

Although never produced as Coca-Cola, perhaps you will recognize this truck as the Pepsi-Cola model Marx eventually produced.

Manufacturer: Marx
Model: Prototype #1129
Composition: Tin
Size: 17½"
Year: 1967
Rarity: 5
Value: N/A
Reference: MX18

Note: "Sample" tag from prototype #1129; LUMAR Company, F.O.B. Glendale, W. Va.

THIS IS ODUE COPY THAT IS ILLEGIBLE AND ALSO CAN NOT BE READ EITHER 1967

2/10/67

GLENDALE DIVISION

SPECIFICATION SHEET

Model No. 8188 M

Cost No. ______

Item COCA COLA TRUCK

Quantity	Part No.	Description		Color
1		1100 Series Cab consisting of		
		PT-1990 Cab Body PT-1991 Cab Top PL-1172 Rad and Grille and PL-1173 Windshield		Plated Radiator and grille
1 1		PT-1992 Chassis 1090 Coca Cola Body consisting of Pt-1647 Bed Fr. and End. PT-1648 Rear End PT-1649 Litho Body		
1	PL-871	Plastic Hand Truck (2 wheel)	Polyeth	Silver
1	PL-83	Wheels	Polyeth	Black
1		PT-1999 Wheel Discs Spoked T.P.L.		
12	PL-31-32	Coke Cases (Hot Stamped and assembled)		
1		Front Axle 187 x 5-5-8		
1		Rear Axle 187x5-5-8		
1	P____	Set 3 decals		
1	P____	6 x 6 Poly bags for cases		

REMARKS:-

DEMONSTRATOR DATA:-

SHIPPING CARTON Size- L........W........H........ Gross........ Tare........ Net........Legal........

SHIPPING CARTON INFORMATION:- 1 Truck packed

Individual Carton-L 7-1 4 W 6-1 8 H 19-3 4 Diagonal 21 Girth 26-3 4 Weight 5-1 2 lbs.

Description corr. box - 2 prints

Master Carton Size-L 18-3 4 W 14-3 4 H 20-1 4 Diagonal 27-1 2 Girth 67 Weight........

Quantity Packed For Shipping Carton 1 2 dozen (6) pcs. per case.

Freight Classification........

Net Weight........ Gross Weight approx. 35 lbs. Tare Weight........ Legal Weight........

Acknowledged by........ Date........

Copy of the Glendale #8188M prospectus sheet for the updated, c. 1967, model #1128 which would have replaced the earlier model #1090. Coca-Cola never licensed it and subsequently it was never produced.

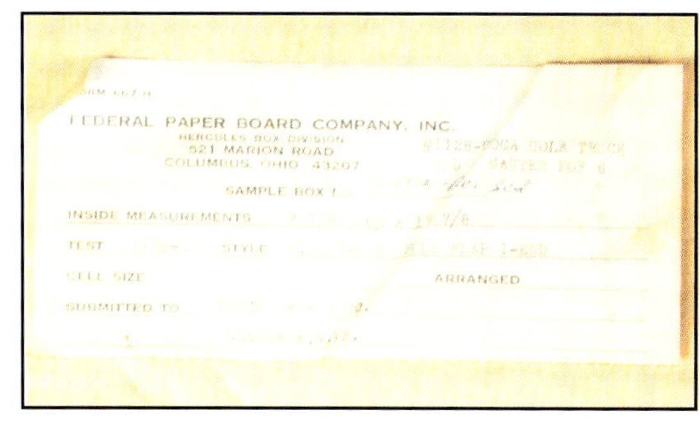

Manufacturer: Marx **Size:** 17½"

Model: Prototype #1128 **Year:** 1967

Composition: Tin **Rarity:** 5

Accessories: 12 cases, hand truck **Value:** N/A **Reference:** MX19

Shown with sample box specifications found on end flap.

HAS ANYONE SEEN THIS TRUCK?

Reference: MX20

Of all the trucks pictured in this book, one in particular has proven to be very mysterious. During our research at the Marx Archives in Miami, Florida, we discovered records indicating that a large-scale plastic truck, approximately 16 inches long (scaled to the 3½" 24-bottle case), had been ordered into production on November 11, 1948. At the time, Louis Marx invested in excess of $21,000 to set-up dies and molds. Attached to the files, referred to as PL 17, 18, and 19, is a photograph of the actual truck, which looks very similar to the 10-inch plastic Marx (MX15) but smaller and without doors. It is baffling because we have never seen this truck and neither has any other collector we have spoken with. By the number of units originally projected in company records, we are inclined to believe that as many as 180,000 should have been produced. Maybe due to early plastic manufacturing problems, the project was scrapped. It is certain that the molds were manufactured because we have seen them. The picture above, courtesy of Marx, is of an actual cab made from these molds. Upon closer examination of the PL 17 Coke truck body spec sheet, we found a small, handwritten note that states: "PL–17–18–19 in Mexico, 1-3-67." Could some lucky Mexican child have played with this truck? If anyone has ever seen or heard of it, please let us know.

NORTH AMERICA

1949–1969

NORTH AMERICA

This chapter includes smaller toy manufacturers, as well as some variations of trucks found in the United States but made specifically for the Canadian or Mexican markets. Among the important smaller United States' manufacturers were Ideal Novelty & Toy Company of Hollis, New York, and Pyro Plastics Corporation of Union City, New Jersey. It is interesting to note that Pyro's name derives from Pyron Park, a section of Union City where the plant was located. Also in this section is the Barclay Manufacturing Company, a company mostly known for its toy soldiers. Strangely, all three of these companies produced only one Coca-Cola vehicle – the rarest being the Ideal truck. Why so few of these Ideals have surfaced is a mystery. We are only aware of one – could it have been a prototype? The age and quality of the Sprite Boy decal and the deep engraving of the logo reinforces our conviction that this is, indeed, an authentic toy.

In this chapter, we also bring you several major Canadian Coke trucks produced by companies including: Lincoln Specialties Limited, Canada's premier toy truck manufacturer, London Toy, Hubley Canada, and Marx Canada. Lincoln Coca-Cola trucks have appeared mostly painted red, but for some reason, they have occasionally surfaced in gray. (Most Marx #99 stake trucks from Canada have also appeared in that color.) These Lincolns also appeared in bilingual versions.The Buvez French version seems to be a little rarer. The London Toy Coca-Cola truck was included in a set of six soft drink trucks including Pepsi-Cola, Orange Crush, 7-UP, and Canada Dry Ginger Ale. The London Toy trucks came in two versions – regular and wind-up. Marx Canada, the largest producer by far, offered trucks identical to those offerred by its American neighbor, except that they were painted in reverse colors – red with yellow accents instead of yellow with red accents in the United States. Another major difference was that the logo, "Where there's Coca-Cola there's hospitality," was indigenous only to Canada. An interesting note was that the cases produced for Marx Canada trucks were a paler yellow color with nearly clear bottles.

We picture two of the rarest Mexican trucks known – the Plasti-Marx made at Marx's Mexican factory and a very rare metal truck that we recognize as a refashioned GAMA. It is obvious that the Mexican toy industry, unable to afford the cost of producing their own tool and dies, often bought obsolete molds from foreign companies such as Tipp & Co. and Gama. The Tipp & Co. molds were used to produce the Lemy VW #323.

Manufacturer: Ideal Novelty & Toy (USA) **Model:** Sprite Boy w/engraved logo **Composition:** Plastic **Size:** 4"
Year: 1949 **Rarity:** 5 **Value:** $600.00–800.00 **Reference:** NA1

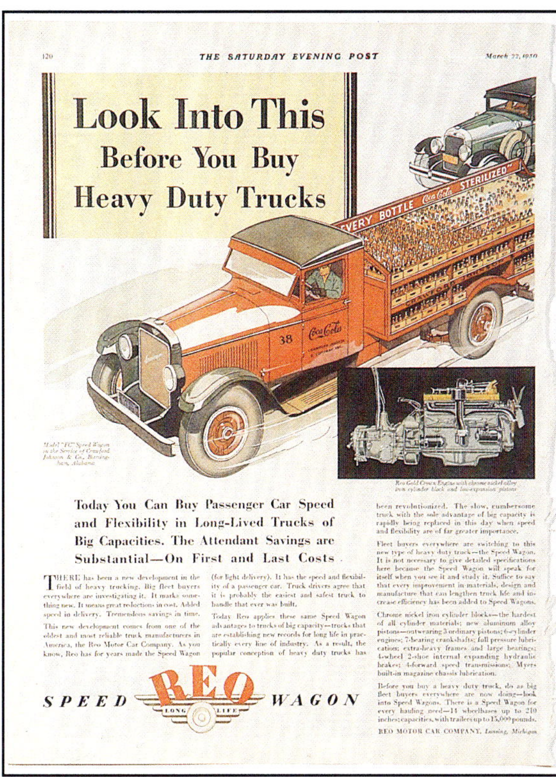

Trade magazine ads showing the *real* delivery trucks that inspired the toy truck industry.

Manufacturer: Pyro Plastic Corp. (USA) **Model:** Unknown **Composition:** Plastic
Size: 5½" **Year:** Early 1950s
Rarity: Smooth rubber wheel -2; knobby plastic tire - 4
Value: Smooth - $75.00–125.00; knobby - $125.00–150.00 **Reference:** NA2

Pyros came in two wheel variations – a smooth rubber and a knobby plastic. See values above.

Manufacturer: Barclay (USA)
Model: #690
Composition: Die-cast metal
Size: Each – 2"
Year: Late 1950s
Rarity: Set – 4; Single truck – 3
Value: Set – $250.00–350.00; Single – $125.00–175.00
Reference: NA3

Photo on right shows Coca-Cola truck #690 from above set.

Manufacturer: Unknown (USA)
Model: Delivery van
Composition: Tin
Size: 1⁹⁄₁₆"
Year: 1950s
Rarity: 5
Value: $600.00–800.00
Reference: NA4

Manufacturer: Unknown (USA) **Model:** Prototype **Composition:** Metal **Size:** 8" **Year:** 1950s **Rarity:** 5 **Value:** N/A
Reference: NA5

Canada

Manufacturer: Marx-Canada **Model:** Delivery **Composition:** Plastic **Accessories:** 2 cases **Size:** 7¼"
Year: 1950s **Rarity:** 5 **Value:** $850.00–950.00 **Reference:** NA6
Note: This model had wooden wheels.

Manufacturer: Marx-Canada **Model:** Chevy style **Composition:** Plastic **Accessories:** 6 pale yellow plastic cases w/clear bottles
Size: 11" **Year:** 1950–54 **Rarity:** 4 **Value:** $1,000.00–1,200.00 **Reference:** NA7
Note: This model only came with wooden wheels and had no doors. It displayed the "Where There's Coca-Cola, There's Hospitality" slogan on the rear.

Reference: NA7

Manufacturer: Marx-Canada **Model:** #991 (all red) **Composition:** Pressed steel **Size:** 20¼" **Year:** 1949–52
Rarity: 5 **Value:** $1,200.00–1,350.00 **Reference:** NA8
Note: Marx-Canada trucks usually displayed the "Where There's Coca-Cola, There's Hospitality" slogan, unique to these Canadian trucks.

Manufacturer: Marx-Canada **Model:** #991 gray cab **Composition:** Pressed steel **Size:** 20¼" **Year:** 1949–52 **Rarity:** 5
Value: $1,200.00–1,350.00 **Reference:** NA9

Manufacturer: Marx-Canada
Composition: Tin
Rarity: 4
Model: #21
Size: 11"
Value: $1,000.00–1,200.00
Accessories: 8 pale yellow plastic cases w/clear bottles, metal hand truck
Year: 1954–56
Reference: NA10

Manufacturer: Lincoln Toy (Canada)
Composition: Pressed steel
Year: Early 1950s
Reference: NA11
Model: #809 stake truck
Size: 16"
Rarity: Gray - 4; Red - 3
Accessories: 12 wooden blocks (sometimes marked Coca-Cola) and cardboard insert with lithographed cases (not shown)
Value: Gray - $800.00–1,000.00; Red - $650.00–850.00

Manufacturer: London Toy copy **Model:** Beverage truck **Composition:** Die-cast metal **Size:** $5\frac{7}{8}$"
Year: 1950s **Rarity:** 3 **Value:** $350.00–450.00 **Reference:** NA12

Manufacturer: London Toy **Model:** Beverage truck, wind-up **Composition:** Die-cast metal **Size:** $5\frac{7}{8}$"
Year: 1950s **Rarity:** 4 **Value:** $650.00–850.00 **Reference:** NA13
Note: These London Toy trucks came in both regular and wind-up versions. The wind-up is the rarest.

Wind-up mechanism of London Toy.

Manufacturer: Hubley-Real Types
Year: 1950s
Model: RT #290
Rarity: 5
Composition: Die-cast metal
Value: $650.00–750.00
Size: 4½"
Reference: NA14

Manufacturer: Hubley-Real Types
Model: RT #290 in original packaging
Composition: Die-cast metal
Size: 4½"
Year: 1950s
Rarity: 5
Value: $1,250.00–1,450.00
Reference: NA15

Mexico

Manufacturer: Plasti-Marx
Model: Chevy style
Composition: Plastic
Accessories: 6 plastic cases
Size: 11"
Year: Late 1950s
Rarity: 4
Value: With Spanish box - $750.00–950.00
Reference: NA16

Note: This truck was sold in Mexico. Although the box was printed in Spanish, the truck and decals were American. This model had no doors.

Manufacturer: Marx
Size: 10"
Reference: NA17
Model: "Snub nose"
Year: Early 1960s
Composition: Plastic
Rarity: 5
Accessories: 6 cases w/black bottles
Value: $1,250.00–1,500.00

Price range reflects trucks in 8.5 condition, not MIB. Same truck with original box could substantially increase its value.

Box indicates that this truck appeared in several beverage versions.

Manufacturer: Gama **Size:** 13" **Reference:** NA18

Model: Refresca Major **Year:** Late 1950s

Composition: Metal **Rarity:** 5

Accessories: 8 red plastic cases, hand truck **Value:** $1,450.00–1,650.00

Note: Although manufacturer cannot be confirmed, obsolete Gama molds were used.

Reference: NA18

EUROPE

1947–1969

EUROPE

Coca-Cola first appeared in Europe between the World Wars, but it was not until the end of World War II that it really became entrenched. The U.S. Army was probably one of Coke's best sales promoters as Coca-Cola was the official drink of the G.I.s who liberated Europe, and who were closely followed by the mobile bottling operations ordered by General Dwight D. Eisenhower. Although subliminal, the identification between the liberators and Coca-Cola was very strong. After the war, most of Europe fell in love with America and her way of life. Every young boy did his best to look and act American, and drinking Coca-Cola became an important part of this process.

It's not surprising to discover that although there were no Coca-Cola toys in Europe before the war, after the war they started appearing in abundance throughout European countries with a suitable industry. Whether the Coca-Cola Company had control over the use of its trademarks as it did in the United States is not clear, but it seems more than likely since in most instances these trucks were usually quite accurate in color, markings and, in most cases, the depiction of actual vehicles.

Interestingly, it was the classic Coca-Cola bottling trucks that fascinated both the manufacturers and their young customers – a kind of vehicle that was new and exciting. Another interesting note is that the color of the toy trucks in Europe was usually a darker shade of yellow, almost orange. The same was true of the real life vehicles. There is no real uniformity of size, style, or even materials among European Coca-Cola toy trucks, although there seem to be certain trends indigenous to different countries.

The only high quality tin-plate Coca-Cola trucks come from Germany where well-established companies like Götz and Son, (a.k.a. Göso), from Fürth, produced wind-up trucks in U.S. Zone - Germany in the late 1940s and early 1950s. Tipp & Co., from Nüremberg, also produced some beautiful models in the mid-1950s. For reasons unknown, no other European country with a tradition of making quality tin-plate toys and vehicles did

Manufacturer: Göso (Germany) **Model:** Wind-up **Composition:** Tin **Accessories:** Unknown **Size:** 7½"
Year: c. 1947–48 **Rarity:** 5 **Value:** $2,500.00–2,800.00 **Reference:** EU1

so. Only some cheap samples emerged from Italy, where something different also took place.

As in the United States, perhaps due to the shortage of raw materials, and/or a weakened wartime economy, the Italians began producing wooden toys. Thanks in particular to a couple of small companies based in Bergamo, not far from Milano, several exciting Coca-Cola vehicles, made mostly of wood, were produced. These companies made beverage trucks, but one of them, Alpia, probably through an arrangement with the Italian Coca-Cola Company, produced several very attractive large size trucks and other vehicles. These Alpia trucks have Bakelite or plastic cabs mounted on wooden frames and bodies. These trucks are very accurate in design and detail, down to the metal license plate with BG-1 for Bergamo. The decals and markings are also quite accurate in their depiction of the standard Coca-Cola markings that were used, including the Sprite Boy and Santa Claus. Actual photographs reveal that these toy trucks were carefully crafted miniature reproductions of the real trucks. These Italian toys deserve a very special place in any Coca-Cola toy truck collection.

Other European countries mainly used plastic, including France, which produced the magnificent Battery-Op (B/O) Camion-Brasseur made by France Jouet, as well as the simpler plastic Sesames with tin accents. Great Britain, by contrast, produced smaller yet accurate die-cast toys from companies, including Budgie and Dinky Toy. Dinky, one of the largest and most established toy companies in England, did not actually produce its salute to Coca-Cola until much later on, in 1965.

Another important English manufacturer is Lesney, which produced the famous #37 Matchbox Coca-Cola truck deriving its name from the simulated numbered matchbox packaging. The Coca-Cola truck was #37 of the 1–75 series. The name "Lesney" comes from a combination of the names of the two founding partners who formed the company in the late 1940s – *Les*lie Smith and Rod*ney* Smith, who, interestingly enough were not related.

John O'Dell, a tool and die maker, rounded off the partnership. In 1982, O'dell would re-emerge as the founder of Lledo, using his name spelled backwards to name the company. Early Matchbox toy boxes read "A MOKO LESNEY" because, until 1959, they were distributed world-wide by *Mo*ses *Ko*hnstam. It is humorous to note how these men played on their names to name the companies as we know them today. By the way, these guys must have really been annoyed by the appearance of the Benbros #49 truck (see page 99) a mirror image of the #37 in almost every detail, including the packaging. Very few of these Benbros #49's have survived. They were produced out of cheaper white metal, prone to cracking and eventual disintegration.

Also included in this chapter are toy trucks manufactured in Holland and Spain along with a very rare example from the former East Germany. (See page 86.) It is ironic that a Communist country would have produced a toy representing the epitome of capitalism. This could have been due to East Germany's need to export products for their economy.

Germany

Manufacturer: Göso
Size: 7½"
Model: Wind-up beverage #426-20
Year: c. 1949 **Rarity:** 4
Composition: Tin
Value: $1,800.00–2,000.00
Accessories: 8 metal cases
Reference: EU2

Close-up of side panel marking on pre-1953 Gösos with "Made in U.S. Zone, Germany" markings.

Right: Rear view of Göso #426-20 Coca-Cola truck.

Manufacturer: Göso
Size: 7½"
Reference: EU3
Model: #430-20, wind-up
Year: 1950–52
Composition: Tin
Rarity: 4
Accessories: 8 metal cases
Value: $1,800.00–2,000.00

Manufacturer: Göso
Size: 8¼"
Reference: EU4
Model: International, wind-up
Year: 1953-55
Note: Sometimes found in red cab variation.
Composition: Tin
Rarity: 4
Accessories: 8 metal or plastic cases
Value: $1,500.00–1,750.00

Manufacturer: Göso
Year: c. 1957

Model: Friction
Rarity: 4

Composition: Tin
Value: $1,400.00–1,600.00

Accessories: 12 red/green plastic cases
Reference: EU5

Size: 13¼"

Reference: EU5

Price range reflects trucks in 8.5 condition, not MIB. Same truck with original box could substantially increase its value.

Reference: EU5

Manufacturer: TCO-Tipp & Co. **Size:** 9"
Model: #323F export version **Year:** c. 1956–66
Composition: Tin **Rarity:** 3
Accessories: 5 plastic cases w/removable clear & red bottles **Value:** $450.00–550.00 **Reference:** EU6

KÖSTLICH Coca-Cola ERFRISCHEND
Coca-Cola
504
Tippco

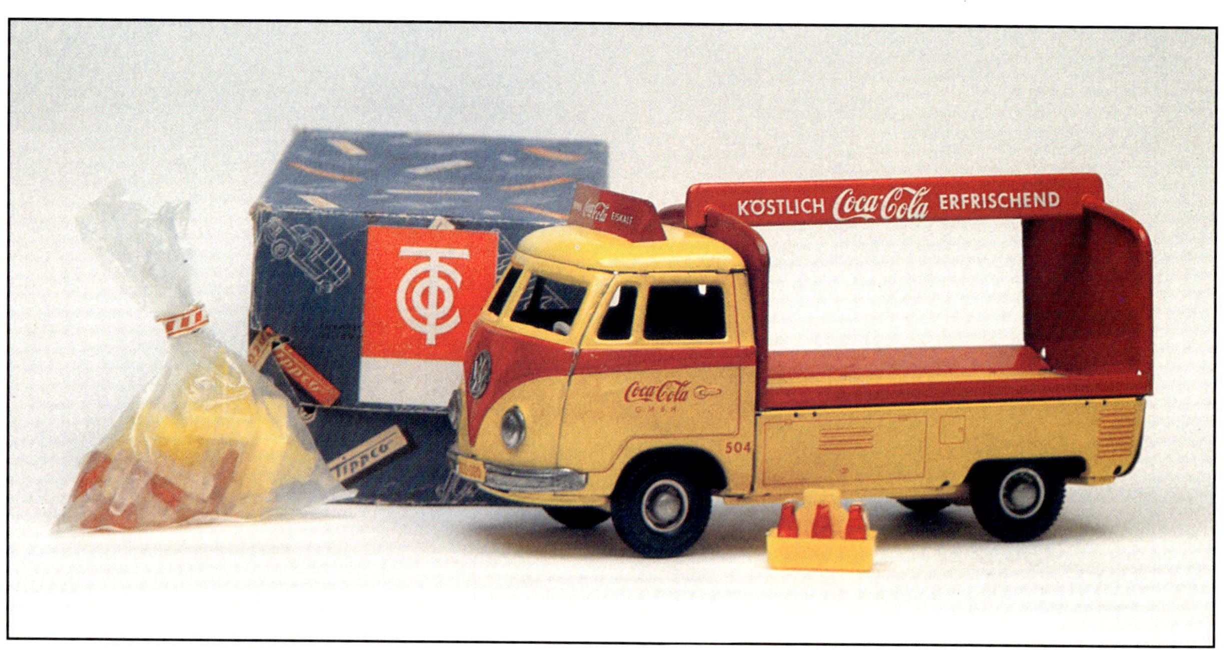

Manufacturer: TCO-Tipp & Co.
Size: 9"
Model: #323F German version
Year: 1956–1966
Composition: Tin
Rarity: 4
Accessories: 5 plastic cases w/removable clear & red bottles
Value: $650.00–750.00
Reference: EU7

Note: Variations include: model w/ driver and a wind-up model.
Rarity: 5
Value: $1,000.00–1,200.00

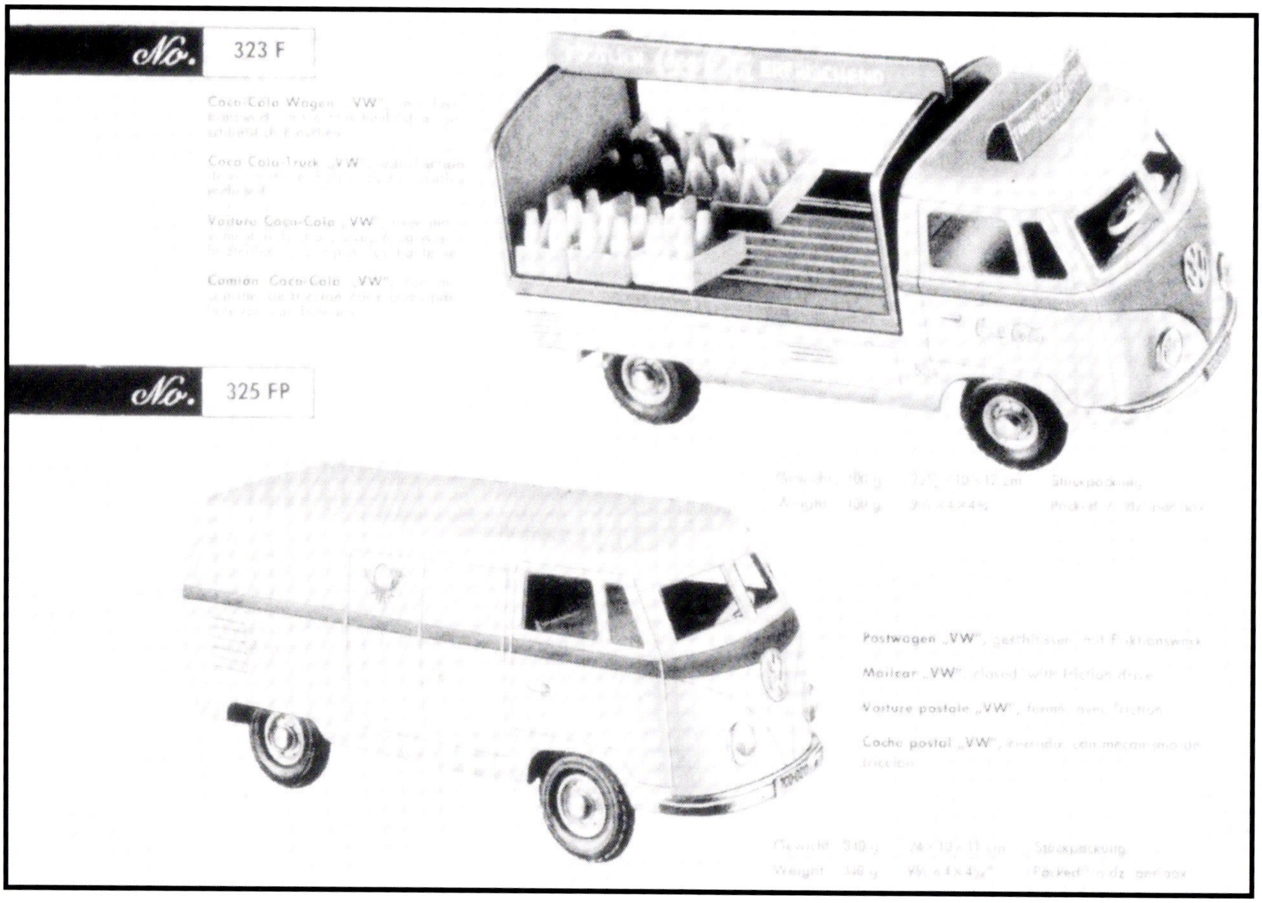
No. 323 F
Coca-Cola-Wagen „VW"
Coca-Cola-Truck „VW"
Voiture Coca-Cola „VW"
Camion Coca-Cola „VW"
No. 325 FP
Postwagen „VW"
Mailcar „VW"
Voiture postale „VW"
Coche postal „VW"

Sales advertisement from German toy catalog, c.1950s.

Manufacturer: TCO-Tipp & Co.
Size: 9"
Value: With color box - $1,450.00–1,650.00

Model: #323F, first version
Year: 1956–58

Composition: Tin
Rarity: With this box - 5
Reference: EU8

Accessories: 6 plastic cases w/removable clear bottles
Note: Absence of G.M.B.H. on door.

Manufacturer: Wiking
Year: Late 1950s–Early 1960s
Composition: Plastic
Rarity: 2–3
Accessories: All models came w/2 strips of 6 brown cases
Value: See individual listings
Size: 3–3½"

Manufacturer: Wiking
Size: 3½"
Value: $125.00–150.00
Model: Ford
Year: 1955
Reference: EU9
Composition: Plastic
Rarity: 2–3

Manufacturer: Wiking
Composition: Plastic
Value: $175.00–225.00
Model: Mercedes, open window
Size: 3¼"
Year: 1963
Rarity: 2–3
Reference: EU10

Manufacturer: Wiking
Size: 3½"
Value: $175.00–225.00
Model: Magirus
Year: 1963
Reference: EU11
Composition: Plastic
Rarity: 2–3

Manufacturer: Wiking
Size: 3"
Value: $200.00–250.00
Model: Bussing
Year: 1963
Reference: EU12
Composition: Plastic
Rarity: 3–4

Manufacturer: Wiking
Composition: Plastic
Value: $125.00–150.00
Note: 2 color variations
Model: Mercedes, closed window
Size: 3¼"
Year: 1955
Rarity: 3
Reference: EU13

Manufacturer: Wiking
Size: 3½"
Value: $125.00–150.00
Model: Ford delivery
Year: 1955
Reference: EU14
Composition: Plastic
Rarity: 3
Note: 2 color variations

Manufacturer: D.D.R. (E.Germany)
Model: Van
Composition: Plastic
Size: Approx. 7"
Year: 1960s
Rarity: 5
Value: $400.00–500.00
Reference: EU15
Note: See text on page 79.

Manufacturer: Gama
Model: Mini Gama VW (left)
Composition: Die-cast
Accessories: 2 white racks of 14 cases each
Size: 4¼"
Year: 1960s
Rarity: 3
Value: $250.00–350.00
Reference: EU16

Manufacturer: Gama
Model: #928 Mini Mod VW, split window (right)
Composition: Die-cast
Accessories: 2 white racks of 14 cases each
Size: 4"
Year: 1950s
Rarity: 3
Value: $350.00–450.00
Reference: EU17

Manufacturer: Schildkröt
Model: Pull toy
Composition: Soft plastic
Accessories: 8 plastic cases
Size: 15"
Year: 1950s
Rarity: 3
Value: $300.00–500.00
Reference: EU18

Made in Germany for Holland market

Manufacturer: Schildkröt
Model: Pull toy
Composition: Soft plastic
Accessories: 8 plastic cases
Size: 15"
Year: 1950s
Rarity: 3
Value: $300.00–400.00
Reference: EU19

Manufacturer: Schildkröt
Model: Pull toy
Composition: Soft plastic
Accessories: 8 plastic cases
Size: 15"
Year: 1960s
Rarity: 3
Value: $300.00–400.00
Reference: EU20

Italy

Manufacturer: Alpia **Model:** Unknown **Composition:** Bakelite, wood, tin **Accessories:** 24 plastic cases w/removable bottles
Size: 17" **Year:** Late 1940s **Rarity:** 5 **Value:** $2,400.00–2,600.00 **Reference:** EU21

Reference: EU21

Manufacturer: Alpia **Model:** Santa Claus **Composition:** Bakelite, wood, tin **Accessories:** 16 cases w/removable bottles
Size: 15½" **Year:** 1950s **Rarity:** 5 **Value:** $2,500.00–2,750.00 **Reference:** EU22

Manufacturer: Alpia **Model:** Unknown **Composition:** Bakelite, wood **Accessories:** 24 cases w/removable bottles
Size: 16½" **Year:** 1950s **Rarity:** 5 **Value:** $2,400.00–2,600.00 **Reference:** EU23

Reference: EU22

Rear view of EU22 and EU23.

Reference: EU23

Manufacturer: Alpia **Model:** Sprite Boy ice cream cart **Composition:** Wood, canvas **Accessories:** 1 case w/24 removable bottles
Size: 7¼" **Year:** Early 1950s **Rarity:** 5 **Value:** $1,750.00–2,000.00 **Reference:** EU24

Manufacturer: Sconosciuta **Model:** Lambretta **Accessories:** 6 plastic ice cream cones
Composition: Plastic **Year:** 1950s **Rarity:** 5 **Value:** $750.00–850.00
Reference: EU25

Manufacturer: Am-Bo(logna)
Model: Unknown
Composition: Tin
Size: 4"
Year: 1950s
Rarity: 5
Value: $750.00–850.00
Reference: EU26

Manufacturer: Am-Bo(logna) **Model:** Articolo #530 **Composition:** Tin **Year:** 1950s
Rarity: 5 **Value:** $1,200.00–1,350.00 **Reference:** EU27
Note: 2 color variations shown

Reference: EU27a

Manufacturer: Am-Bo(logna) **Model:** Super Luxuosa Articolo #530 (set of 12) **Composition:** Tin **Year:** 1950s
Rarity: 5 **Value:** $1,650.00–1,850.00 **Reference:** EU28

Manufacturer: Am-Bo(logna)
Year: 1960s
Reference: EU29

Model: Mignon-Lux (set of 12)
Rarity: 4
Note: Coca-Cola truck in this box has knobby tires.

Composition: Tin
Value: $800.00–1,000.00

Manufacturer: Am-Bo(logna)
Model: Coca-Cola truck (3 variations)
Composition: Tin
Size: 4"
Year: 1960s
Rarity: Knobby and smooth tires - 3
Tin wheels - 5
Value: Knobby tires - $350.00–450.00
Smooth tires - $500.00–600.00
Tin wheels - $750.00–850.00
Reference: EU30, 31, 32

Manufacturer: Am-Bo(logna) **Model:** Mignon #413-12 (set of 12) **Composition:** Tin **Year:** 1960s **Rarity:** 4
Value: $800.00–1,000.00 **Reference:** EU33

Note: These two sets have Coca-Cola trucks with knobby tires.

Manufacturer: Am-Bo(logna) **Model:** Mignon #400-6 (set of 12) **Composition:** Tin **Year:** 1960s **Rarity:** 4
Value: $800.00–1,000.00 **Reference:** EU34

Manufacturer: National Toys
Size: 12½"
Reference: EU35
Model: Fiat
Year: 1960s
Composition: Plastic
Rarity: 4
Accessories: 8 cases w/12 bottles
Value: $650.00–750.00

Manufacturer: Unknown
Size: 10"
Reference: EU36
Model: Fiat delivery
Year: 1960s
Composition: Plastic
Rarity: 4
Accessories: 6 cases w/12 bottles
Value: $350.00–450.00

United Kingdom

Original point of purchase display, c.1950s. Notice the 49¢ price.

Manufacturer: Lesney **Model:** Matchbox #37 **Composition:** Die-cast **Size:** 2¼" **Year:** 1950s–60s
Rarity: 2 **Value:** See individual listings **Reference:** EU37

Manufacturer: Lesney
Composition: Die-cast
Rarity: 2
Reference: EU38
Model: Matchbox #37A (uneven load)
Size: 2¼"
Year: c. 1950s
Value: $125.00–135.00

Manufacturer: Lesney
Composition: Die-cast
Rarity: 1
Model: Matchbox #37A (even load, no base plate)
Size: 2¼"
Year: Late 1950s
Value: $75.00–85.00
Reference: EU39

Manufacturer: Lesney
Composition: Die-cast
Rarity: 1
Model: Matchbox #37B (even load w/base plate)
Size: 2¼"
Year: 1960s
Value: $60.00–75.00
Reference: EU40

Manufacturer: Lesney
Composition: Die-cast
Rarity: 1
Reference: EU41
Model: Matchbox #37C (black wheels)
Size: 2¼"
Year: Late 1960s
Value: $50.00–60.00

Manufacturer: Benbros
Composition: Die-cast
Rarity: 5
Reference: EU42
Model: Mighty Midget #49
Size: 2¼"
Year: 1950s
Value: $200.00–250.00
Note: See text on page 79.

Manufacturer: Budgie **Model:** Van **Composition:** Die-cast **Accessories:** 11 brown plastic cases **Size:** 5¼"
Year: Late 1950s **Rarity:** 3 **Value:** $300.00–400.00 **Reference:** EU43

Rear view
Reference: EU43

Price range reflects trucks in 8.5 condition, not MIB. Same truck with original box could substantially increase its value.

Manufacturer: Dinky-Toy
Model: Bedford #402
Composition: Die-cast
Accessories: 6 strips of plastic cases
Size: 4½"
Year: 1965–69
Rarity: 4
Value: $325.00–425.00
Reference: EU44

Manufacturer: Corgi
Model: Lunch van (with spin-around cook)
Composition: Die-cast
Size: 3¾"
Year: 1966–69
Rarity: 2
Value: $100.00–150.00
Reference: EU45

Spain

Manufacturer: Anguplas **Model:** Ebro #11 **Composition:** Plastic **Size:** 2¾" **Year:** 1950s
Rarity: 4 **Value:** $200.00–300.00 **Reference:** EU46

Manufacturer: Anguplas **Model:** Seat #1400
Composition: Plastic **Size:** 2¼" **Year:** Late 1950s
Rarity: 4 **Value:** $175.00–250.00
Reference: EU47

Manufacturer: Anguplas **Model:** Autobus
Composition: Plastic **Size:** 3" **Year:** Late 1950s
Rarity: 3 **Value:** $125.00–175.00
Reference: EU48

Manufacturer: Gozan **Model:** Tigré 2000 **Composition:** Metal, plastic **Accessories:** 24 cases **Size:** 12¾"
Year: 1960s **Rarity:** 3 **Value:** $250.00–350.00 **Reference:** EU49

Manufacturer: Gozan **Model:** Tigré 317 **Composition:** Metal, plastic **Accessories:** 24 cases **Size:** 12¼"
Year: 1960s **Rarity:** 3 **Value:** $250.00–350.00 **Reference:** EU50

Manufacturer: Gozan **Year:** 1960s

Model: Tigré 317 variation **Rarity:** 3

Composition: Metal, plastic **Value:** $250.00–350.00

Accessories: 24 cases **Reference:** EU51

Size: 12¼"

France

Manufacturer: Havas **Value:** $400.00–500.00

Model: Renault cut-out **Reference:** EU52

Composition: Cardboard

Size: 15" x 11"

Year: 1954

Rarity: 5

Reference: EU53

Manufacturer: France Jouets **Model:** Camion Brasseur remote control **Composition:** Plastic
Accessories: 10 plastic red or yellow cases w/green bottles, hand truck **Size:** 10½" **Year:** 1960s
Rarity: 4 **Value:** $1,250.00–1,500.00 **Reference:** EU53

Manufacturer: Sesame **Model:** Estafette, friction power (box set of 6) **Composition:** Tin, plastic
Size: 3¼" **Year:** 1960s **Rarity:** 4 **Value:** $750.00–850.00
Reference: EU54

Manufacturer: Sesame (3 variations) **Model:** Estafette (no power) **Composition:** Tin, plastic **Size:** 3¼" **Year:** 1960s
Rarity: 3 **Value:** $150.00–250.00 each **Reference:** EU55

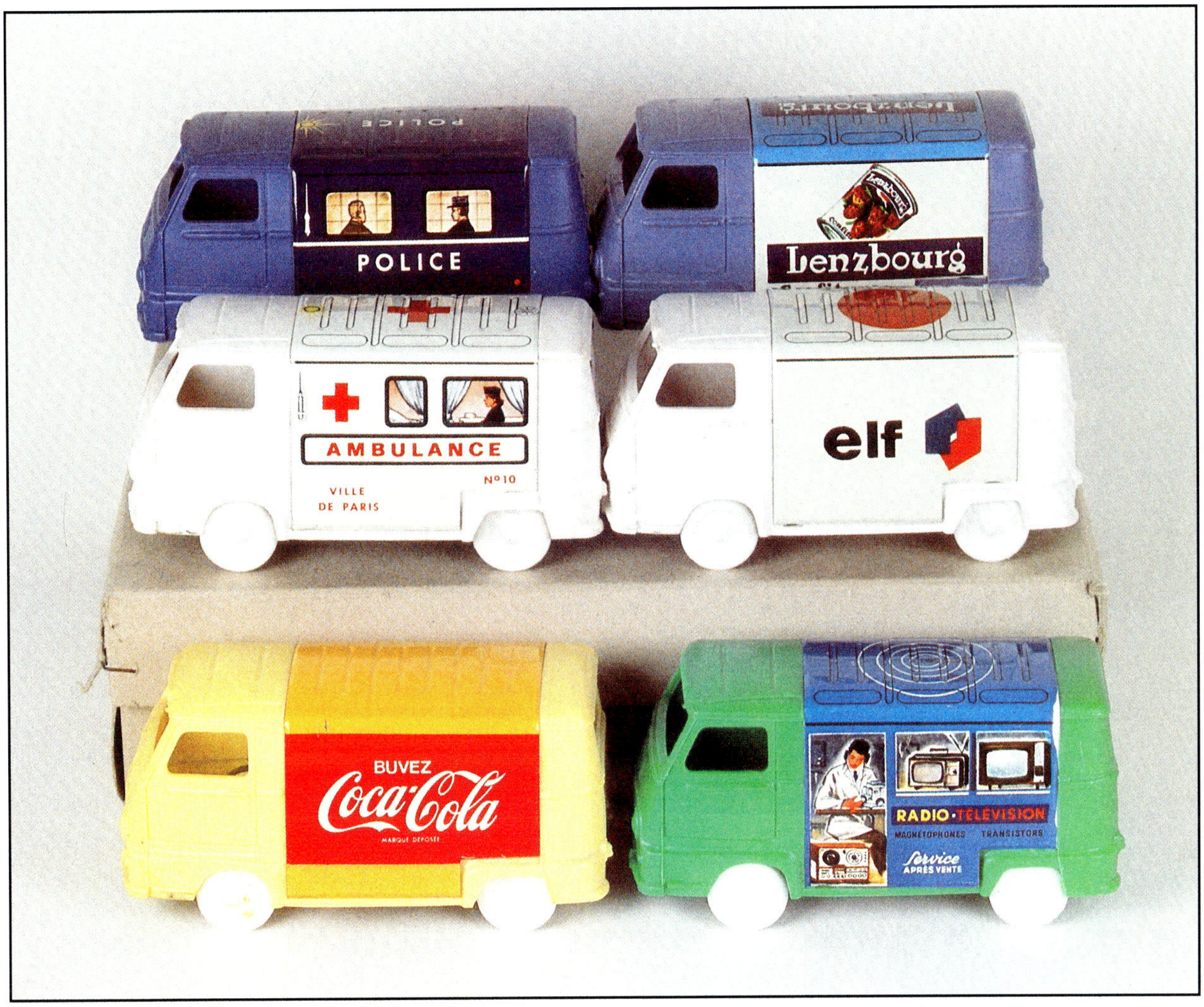

Manufacturer: Sesame **Model:** Estafette, large scale (box set of 6) **Composition:** Tin, plastic **Size:** Each - 5" **Year:** 1960s
Rarity: 5 **Value:** Set - $1,000.00–1,200.00; Single - $400.00–500.00 **Reference:** EU56

Sesame Estafette size comparison. The large scale model is *extremely* rare.

Manufacturer: France Jouets
Composition: Die-cast
Year: Early 1960s
Value: $250.00–350.00
Model: Stradair
Size: 4⅛"
Rarity: 3
Reference: EU57

Manufacturer: France Jouets
Composition: Die-cast
Year: Mid 1960s
Value: $400.00–500.00
Model: Berliet
Size: 3¾"
Rarity: 5
Reference: EU58

JAPAN and ASIA

1950s–1960s

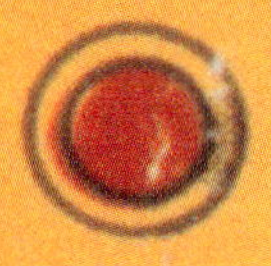

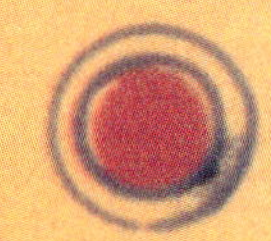

JAPAN AND ASIA

The Japanese toy industry emerged like a phoenix from the ashes of World War II. In the late 1940s, Japan was laden with a surplus of military factories filled with obsolete war machinery and excess materials. Japanese industry, facing a defeated, crumbled economy, needed to generate employment and income for its people. Since the machinery and assembly lines which had been used for the production of war materials were still in place, the transition to peacetime manufacturing was a simple process. Over the next few years, several hundred toy factories, mostly concentrated near Tokyo and Osaka, went into production. Companies, including Marusan, Cragstan, Taiyo, KTS, Haji, and Linemar, went about the task of manufacturing some of the most interesting, unique, and creative Coca-Cola toy trucks ever produced. Louis Marx & Company, with its Japanese operation, Linemar, was a forerunner in educating the fledging Japanese toy industry in the tin lithography process. They also manufactured an industry standard, the Friction Motor. Linemar produced several models, often with variations. It is not unusual to find a certain model with cab, wheel or grille differences.

By the early 1960s, Japan had finely tuned its toy industry and had now become the world leader, exporting nearly 100 million dollars worth of toys a year. More than 50% of these were exported to the United States. The fact that these very attractive and colorful toys were reasonably priced certainly enhanced their popularity in the United States' market. Unfortunately, like their United States and European counterparts, the Japanese tin toy industry would eventually fall victim to the emergence of plastics and the rising cost of labor. The Japanese toy industry was eventually consumed by its neighbors. Production lines could be manned by cheap labor in places like Hong Kong, Korea, and eventually, China. This chapter includes most of the wonderfully innovative and rare examples of Coca-Cola toy trucks produced in Japan before 1970, including some very rare boxed sets.

Manufacturer: "H" **Model:** GMC, friction power **Composition:** Tin **Size:** 5½" **Year:** Early 1950s **Rarity:** 5
Value: $1,250.00–1,500.00 **Reference:** JA1

Manufacturer: ASC
Model: Transport trailer, friction power
Composition: Tin
Size: 8¾"
Year: Early 1950s
Rarity: 4
Value: $800.00–1,000.00
Reference: JA2

Reference: JA3

Manufacturer: ASC
Model: Transport trailer, friction power (Boxed set of 6)
Composition: Tin
Size: Individual: 8¾"
Year: Early 1950s
Rarity: Set - 5; Individual - 4
Value: Set - $1,600.00–1,800.00
Individual - $800.00–1,000.00
Reference: JA3-4

Manufacturer: Marusan (San)
Model: #3431, friction power
Composition: Tin
Accessories: 2 removable tin bottle racks
Size: 8"
Year: 1956–57
Rarity: 3
Value: $750.00–850.00
Reference: JA5

Reference: JA5

Note: License plates indicate that this truck (JA5) was produced in two versions – 1956 and 1957.

Price range reflects trucks in 8.5 condition, not MIB. Same truck with original box could substantially increase its value.

Manufacturer: Rosko **Model:** Beverage delivery, friction power **Composition:** Tin **Size:** 8" **Year:** 1950s **Rarity:** 3
Value: $650.00–750.00 **Reference:** JA6

Manufacturer: Haji **Model:** Friction power **Composition:** Tin **Size:** 4¼" **Year:** 1950s **Rarity:** 4
Value: $650.00–750.00 **Reference:** JA7

Manufacturer: KTS **Model:** Corvair, friction power **Composition:** Tin w/cardboard insert
Size: 8" **Year:** 1950s **Rarity:** With this box - 5; with cardboard insert - 4
Value: With color box - $1,750.00–2,000.00; Without box, but with cardboard insert - $800.00–1,000.00
Reference: JA8

Manufacturer: KTS
Model: Corvair pick-up, friction power, no inserts
Composition: Tin
Size: 8"
Year: 1950s
Rarity: 3
Value: $550.00–650.00
Reference: JA9
Note: Corvair pick-up sometimes appears in a white roof variation, same value.

Manufacturer: TT
Model: Van (bread truck), friction power (Boxed set of 12)
Composition: Tin
Year: 1950s
Rarity: Set - 4; Single: 1
Value: Set - $650.00–750.00; Single - $125.00–150.00
Reference: JA10

View of the 12 different vans in their box.
Reference: JA10

Model on left with rear opening door is rare.
Rarity: 5 **Value:** $250.00–350.00 **Reference:** JA10a

Manufacturer: Unknown
Rarity: 5
Model: Santa tandem trailer, friction power
Value: $2,000.00–2,250.00
Composition: Tin
Reference: JA11
Size: 12¼"
Year: 1957

Rear view of Santa tandem trailer.

Reference: JA11

Manufacturer: Linemar
Model: Squash cab, friction power
Set of 6 (Pepsi truck missing)
Composition: Tin
Size: Each - 3"
Year: 1950s
Rarity: Set - 3; Single - 1
Value: Set - $375.00–475.00;
Single - $175.00–225.00
Reference: JA12

Manufacturer: Linemar
Model: Friction power w/paper label
Composition: Tin
Size: 3½"
Year: 1950s
Rarity: Left - 4; Right - 2
Value: Left - $250.00–350.00;
Right - $175.00–225.00
Reference: JA13, JA14
Note: JA13 also comes in a gold cab variation.
Rarity: 5 **Value:** $275.00–375.00

Manufacturer: Linemar
Model: Friction power
Composition: Tin
Size: 5"
Year: 1950s
Rarity: 3
Value: $450.00–550.00
Reference: JA15
Note: Caution! This truck is often marked Cola Cola. Prices reflect only trucks marked *Coca*-Cola.

Manufacturer: Cragstan
Model: Tiny Giant, friction power (Boxed set of 12)
Composition: Yellow cab - Tin; Red cab - Tin, plastic
Size: Each - 4"
Year: Late 1950s–early 1960s
Rarity: Set - 4; Individual - 2
Value: Set - $500.00–600.00; Individual - $125.00–150.00
Reference: JA16, JA17, JA18, JA19

Reference: JA16

Reference: JA17

Reference: JA18

Reference: JA19

Manufacturer: Endoh
Model: Volkswagen, friction power
Composition: Tin
Size: 4"
Year: 1950s
Rarity: 3
Value: $225.00–325.00
Reference: JA20

Manufacturer: Endoh
Model: Volkswagen, very rare variation w/enclosed body and logo on roof
Composition: Tin
Size: 4"
Year: 1950s
Rarity: 5
Value: $650.00–750.00
Reference: JA21

Manufacturer: Taiyo
Model: Volkswagen, friction power
Composition: Tin
Size: 8½"
Year: 1960s
Rarity: 2
Value: $250.00–350.00
Note: This VW also comes in a blue windshield variation; same value.
Reference: JA22

Manufacturer: Taiyo **Model:** Soda car series, friction power **Composition:** Tin **Size:** 8½" **Year:** 1960s **Rarity:** 2
Value: \$225.00–325.00 **Reference:** JA23, JA23a

These soda cars come in two variations. Car on left has moving eyes (JA23); the one on the right has stationary eyes (JA23a). Value is the same.

Manufacturer: Unknown **Model:** Station wagon, friction power (boxed set of 6) **Composition:** Tin **Size:** Each - 5½" **Year:** 1960s

Rarity: 5 **Value:** Set - $2,250.00–2,500.00; Individual (in packaging) - $1,400.00–1,500.00 **Reference:** JA24

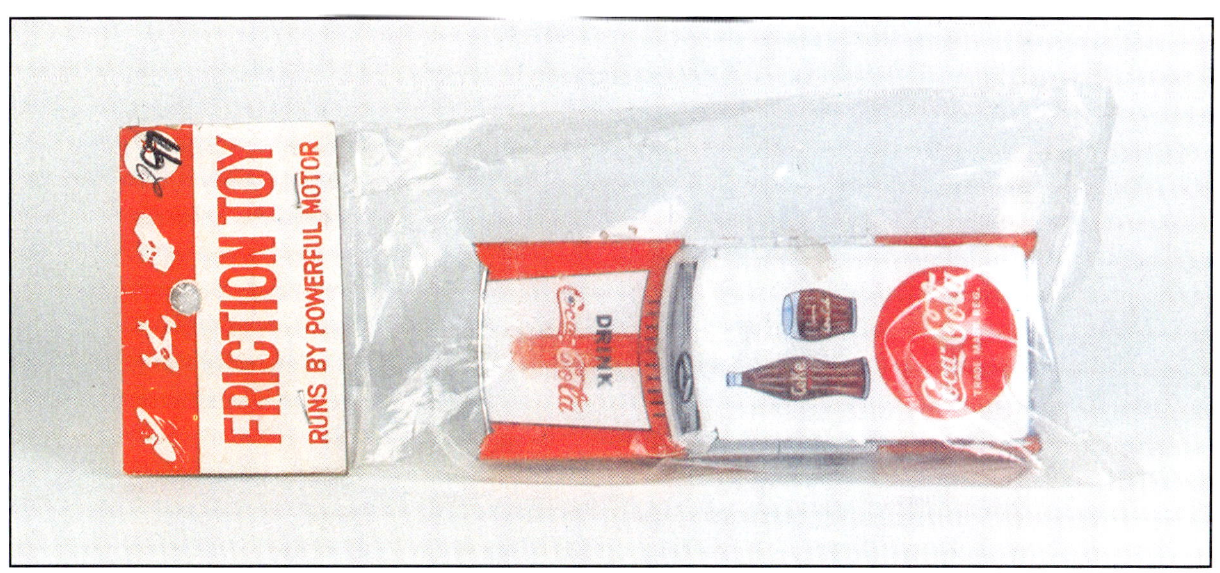

Manufacturer: Unknown **Model:** Station wagon, friction power **Composition:** Tin **Size:** 5½" **Year:** 1960s **Rarity:** 5
Value: $1,250.00–1,450.00 **Reference:** JA25

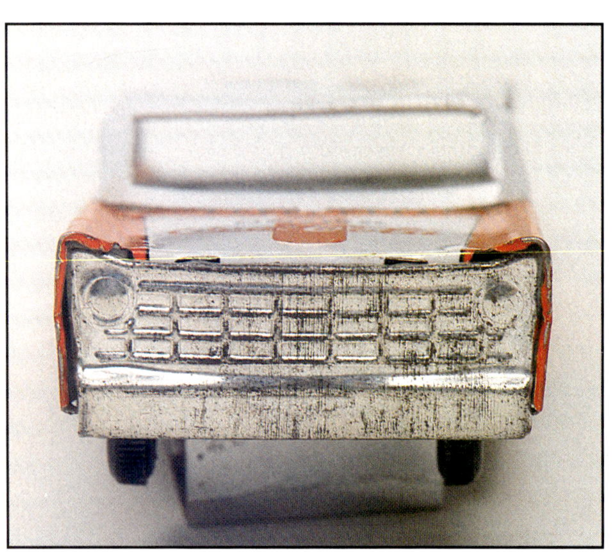

Front and rear view of friction-powered station wagon.

Manufacturer: Taiyo **Model:** Ford Zest car, friction power (2 variations shown) **Composition:** Tin
Size: 9" **Year:** 1960s **Rarity:** 2 **Value:** $225.00–275.00
Reference: JA26

Manufacturer: Taiyo or Dott. G.R. **Model:** Ford taxi, friction power **Composition:** Tin **Size:** 9"
Year: 1960s **Rarity:** 3 **Value:** $250.00–350.00
Reference: JA27

Battery-operated route truck in action. **Reference:** JA28

Manufacturer: Sanyo
Model: Non-stop route truck, battery operated (yellow version)
Composition: Tin
Size: 12½"
Year: 1960s
Rarity: 2
Value: $250.00–350.00
Reference: JA28
Note: Featured on previous page.

Rear graphics on battery-operated route trucks.
Reference: JA28, JA29.

The yellow version of this truck came in two box variations. Coca-Cola is printed in block letters on this box; script on the box shown above.

Manufacturer: Sanyo
Model: Non-stop route truck, battery operated (red version)
Composition: Tin
Size: 12½"
Year: 1960s
Rarity: 4
Value: $350.00–450.00
Reference: JA29

Price range reflects trucks in 8.5 condition, not MIB. Same truck with original box could substantially increase its value.

Manufacturer: N
Model: Ice cream truck soda car; wind-up
Composition: Tin
Size: 4½"
Year: 1960s
Rarity: 4
Value: $750.00–850.00
Reference: JA30

Reference: JA30

Manufacturer: Unknown (Japan)
Model: Friction power
Composition: Tin
Size: 5"
Year: 1960s
Rarity: 5
Value: $1,200.00–1,400.00
Reference: JA31

Manufacturer: Unknown (Japan)
Model: Friction power
Composition: Tin, plastic
Size: 3½"
Year: 1960s
Rarity: 5
Value: $350.00–450.00
Reference: JA32
Note: Also came in red cab variation.

Manufacturer: Unknown (Japan)
Model: Friction power
Composition: Tin, plastic
Size: 6"
Year: 1960s
Rarity: 5
Value: $650.00–750.00
Reference: JA33

Manufacturer: Unknown (Japan) **Model:** City bus, friction power **Composition:** Tin **Size:** 17¾" **Year:** 1960s **Rarity:** 5
Value: $1,500.00–1,600.00 **Reference:** JA34

Manufacturer: M.A. (Pakistan) **Model:** Service bus, friction power **Composition:** Tin **Size:** 7¼" **Year:** 1960s **Rarity:** 4
Value: $1,000.00–1,200.00 **Reference:** JA35

Manufacturer: Blue Box (Hong Kong)
Model: "Super smooth" friction power (Dinky-Toy copy)
Composition: Plastic
Accessories: 6 strips of cases
Size: 6"
Year: 1970s
Rarity: 4
Value: $175.00–225.00
Reference: JA36

Manufacturer: NFIC (Hong Kong)
Model: Bedford, friction power (Dinky-Toy copy)
Composition: Plastic
Accessories: 6 strips of cases
Size: 6¼"
Year: 1980s
Rarity: 4
Value: $175.00–225.00
Reference: JA37

Manufacturer: NFIC (Hong Kong)
Model: #3079 Bedford, friction power (Dinky-Toy copy)
Composition: Plastic
Accessories: 6 strips of cases
Size: 6¼"
Year: 1980s
Rarity: 3
Value: $80.00–100.00
Reference: JA38

Manufacturer: NFIC (Hong Kong)
Model: Delivery, friction power
Composition: Plastic
Accessories: 6 strips of cases
Size: 5¼"
Year: 1980s
Rarity: 4–5
Value: $225.00–325.00
Reference: JA39

Manufacturer: Unknown
Model: Advertising trucks
Composition: Plastic w/paper inserts
Size: Each - 4½"
Year: 1960s
Rarity: 4
Value: Set - $250.00–350.00;
Individual - $125.00–175.00
Reference: JA40

Manufacturer: FE
Model: #620/3 Flip-O-Matic Little People, spring lever activated
Composition: Plastic
Size: 4½"
Year: 1964
Rarity: 4
Value: $100.00–125.00
Reference: JA41
Note: This model is the original of this style. It is quite rare. In the 1970s, Durham Industries reintrodued it as the "Zip-Along." It appeared in white or yellow with lime green hubs and had the familiar spring action. It also had the new Coca-Cola logo. Production continued into the 1980s, but it was called the "Roll-Along" because the spring action was discontinued. See page 152.

MODERN ERA/NEW LOGO

1970–1986

MODERN ERA/NEW LOGO

In 1970, the Coca-Cola Company initiated a multimillion dollar image refurbishing campaign, which introduced the new "Dynamic Contour" logo – a twisting white ribbon under the Coca-Cola and Coke trademarks. The majority of all Coca-Cola toy trucks ever produced have "the Wave" as it is sometimes referred to. This new logo often serves as a cut-off point to the purist collector. New and ardent collectors, however, relish the myriad of trucks marketed since this date, as they sometimes cannot afford the high prices that the vintage toys command. The following chapter is by far the largest in the book. It includes all Coca-Cola toy trucks that have this new logo. Beginning in 1970 and continuing today, this logo appears on toy trucks produced all over the world.

This chapter begins with the new generation Buddy "L" trucks, now manufactured in Japan under strict U.S. specifications, and continues with some of the popular trucks that we have come to enjoy over the years. Included are some rare and beautiful ones from diverse and exotic countries such as Singapore, India, Australia, China, Hungary, Bulgaria, Argentina, and Brazil. Some of the manufacturers represented in this chapter include Tomica, Siku, Yaxon, Sanson, Corgi, Matchbox, Mattel, Winross, and Nylint, as well as many others. Although we have strived to bring to you the most complete selection of trucks with the new logo, we certainly could have missed some variations or models. However, this chapter includes most of the important ones!

Manufacturer: Buddy "L"
Model: #5117 chrome hub; #5117C white hub
Composition: Steel
Accessories: Chrome hub – 10 mini cases, 1 hand truck
White hub – 5 cases, 1 hand truck
Size: 9⅛"
Year: Early 1970s
Rarity: 1
Value: \$50.00–75.00
Reference: NL1 (left), NL1a (right)
Note: This was the first Coca-Cola truck with the new logo to appear on the market. Interestingly, it was sold concurrently with the model #5426 yellow Buddy "L," shown on page 40, for several years.

Manufacturer: Buddy "L" **Model:** #4999 Pop Art Buggies (set of 4) **Composition:** Plastic **Year:** 1971–74
Rarity: Set - 4; Individual - 1 **Value:** Set - $250.00–350.00; Individual - $40.00–50.00 **Reference:** NL2

Two of the different sets that were available. **Reference:** NL2, NL2a

Reference: NL3, NL3a, NL3b

Reference: NL3b

Manufacturer: Buddy "L"
Model: Brute trailer; #4973+ Brute set; #4973D Brute set (Sets contain 5 pieces)
Composition: Steel
Accessories: Trailer - 5 cases, 1 hand truck; Sets - Fork lift, steel loading ramp, 4 cases
Size: Trailer - 10¾"
Year: 1981
Rarity: 1
Value: Trailer - $20.00–30.00; #4973+ - $65.00–85.00; #4973D - $50.00–75.00
Reference: NL3 (top) , NL3a (bottom left), NL3b (bottom right)

Reference: NL3a

Manufacturer: Buddy "L" **Model:** #4969E closed metal sides (6-piece set) **Composition:** Steel **Accessories:** Fork lift, 3 cases
Year: Early 1980s **Rarity:** 3 **Value:** $75.00–85.00 **Reference:** NL4
Note: The closed metal sides version of the truck shown in this set is by far the rarest of this style. Truck by itself - $50.00–60.00.

Manufacturer: Buddy "L" **Model:** #4973E (7-piece set) **Composition:** Steel **Accessories:** Fork lift, 3 cases, steel loading ramp
Year: 1980s **Rarity:** 1 **Value:** $50.00–65.00 **Reference:** NL5

Manufacturer: Buddy "L"
Model: #666H (15-piece set)
Composition: Steel
Accessories: 10 cases, 2 hand trucks, figure
Year: 1980s
Rarity: 1
Value: $35.00–45.00
Reference: NL6

Right: Four different boxes picturing different set variations.

Manufacturer: Buddy "L"
Year: 1980s
Model: #666J (15-piece set)
Rarity: 1
Composition: Steel
Value: $25.00–35.00
Accessories: 10 cases, 2 hand trucks, figure
Reference: NL7

Manufacturer: Buddy "L"
Size: Each - 4¾"
Reference: NL8, NL9, NL10

Model: Left to right: #420C, #420, #4201I
Year: 1980s

Composition: Steel
Rarity: 1

Accessories: 2 plastic cases
Value: #420C - $15.00–20.00; #420 - $10.00–15.00; #4201I - $8.00–12.00

Manufacturer: Buddy "L"
Year: 1980s

Model: #5215H
Rarity: 1

Composition: Steel
Value: $25.00–35.00

Accessories: 4 cases
Reference: NL11

Size: Each - 7½"

Manufacturer: Buddy "L"
Model: #4885E Lil Brutes
Composition: Steel
Accessories: 1 cases
Size: Each - 4¾"
Year: 1981
Rarity: 3
Value: $35.00–45.00
Reference: NL12

Manufacturer: Buddy "L"
Model: #591G Mack trailer
Composition: Steel
Accessories: 5 cases, hand truck
Size: 10½"
Year: 1980s
Rarity: 1
Value: $20.00–25.00
Reference: NL13
Note: Also appeared in "Coke Is It!" and "It's The Real Thing" versions.

Manufacturer: Buddy "L"
Model: #5270J (with or without sunroof and with several different slogans), see NL13
Composition: Steel
Accessories: 8 cases, 1 Coke machine
Size: 14"
Year: 1980s
Rarity: 1
Value: $25.00–35.00
Reference: NL14

These are just two examples of the widespread international distribution of Buddy "L" trucks. We have seen similar trucks from many different nations.

Manufacturer: Aiwa/Buddy "L" (Japan) **Model:** #591-1350 **Composition:** Steel
Accessories: 5 cases, 1 hand truck **Size:** 10½" **Year:** 1980s **Rarity:** 3
Value: $30.00–40.00 **Reference:** NL15

Manufacturer: Rollet-Buddy "L" (France) **Model:** #6-5270 **Composition:** Steel **Accessories:** 8 cases, Coke machine
Size: 14" **Year:** 1980s **Rarity:** 3 **Value:** $35.00–45.00 **Reference:** NL16

Manufacturer: Taiyo (Japan) **Model:** Big Wheel remote control w/blinking lights **Composition:** Plastic, tin **Size:** 10¼"
Year: 1970s **Rarity:** 5 **Value:** $350.00–450.00 **Reference:** NL17

Reference: NL17

Manufacturer: Taiyo (Japan) **Model:** Bump 'n Go Big Wheel, battery-operated w/blinking lights **Composition:** Plastic, tin
Size: 10¼" **Year:** 1970s **Rarity:** 1 **Value:** $80.00–100.00
Reference: NL18 **Note:** This truck appears in 3 versions – Coca-Cola Bottling Co., New York Bottling Co., and Atlanta Bottling Co. (same value).

Price range reflects trucks in 8.5 condition, not MIB. Same truck with original box could substantially increase its value.

Manufacturer: S.T. (Japan) **Model:** Double decker bus, friction power **Composition:** Tin **Size:** 17¾"
Year: 1970s **Rarity:** 3 **Value:** $250.00–350.00
Reference: NL19

Manufacturer: TPS (Japan) **Model:** Jumbo trailer, friction power **Composition:** Tin, plastic **Size:** 16"
Year: 1970s **Rarity:** 3 **Value:** $225.00–325.00 **Reference:** NL20

Manufacturer: Daiya (Japan) **Model:** Isuzu cola truck w/clockwork cargo doors, friction power **Composition:** Plastic
Accessories: 9 cases, hand truck **Size:** 12" **Year:** 1980s **Rarity:** 3
Value: $150.00–250.00 **Reference:** NL21

Left: Clock work mechanism that releases rear doors of Isuzu cola truck.

Below: Box shows instructions for swinging clockwork doors on truck.

Manufacturer: T.T. (Japan)
Model: Van truck, friction power
Composition: Tin, plastic
Size: 3¾"
Year: 1970s
Rarity: 3
Value: Set - $175.00–275.00; Individual - $25.00–35.00
Reference: NL22
Note: Misspelled "REFRESING" variation (see last on left).
Rarity: 3 **Value:** $35.00–45.00.

Unusual red cab variation.
Value: $40.00–50.00. **Reference:** NL22a

Manufacturer: Asahi (Japan)
Model: #7672 van truck, friction power
Composition: Plastic, tin
Size: 4¾"
Year: 1970s
Rarity: 2
Value: $65.00–85.00
Reference: NL23

Manufacturer: Time (Japan)
Model: Town car, wind up (Sct of 24)
Composition: Tin
Size: Fach - 2¾"
Year: 1980s
Rarity: Set - 5; Individual - 4
Value: Set - $400.00–600.00; Individual - $225.00–275.00
Reference: NL24 (above), NL25 (right)

Manufacturer: Unknown (Japan)
Model: Van line, wind-up
Composition: Tin
Size: 3$\frac{5}{16}$"
Year: 1980s
Rarity: 3
Value: $175.00–225.00
Reference: NL26

Manufacturer: WM (Japan, Hong Kong, Taiwan)
Model: Super Mini-Series, friction power
Composition: Tin
Size: Each - 1½"
Year: 1980s
Rarity: Japan - 3; Hong Kong - 2; Taiwan - 1
Value: Set: Japan - $150.00–250.00; Hong Kong - $125.00–175.00; Taiwan - $100.00–125.00
Individual: Japan - $25.00–35.00; Hong Kong - $15.00–25.00; Taiwan - $10.00–20.00
Reference: NL27 (Japan), NL27a (Hong Kong), NL27b (Taiwan)

Note: Red cab variation (second from left)
Manufacturer: Agglo (Hong Kong)
Rarity: 4+
Value: $75.00–85.00
Reference: NL27c

Manufacturer: Yone (Japan), T.T. (Japan) **Model:** Racing cars (Yone - friction power, T.T. - wind-up) **Composition:** Tin **Size:** Each - 3"
Year: 1980s **Rarity:** 1 **Value:** Blue/yellow cars (Yone) - $15.00–25.00. Red/white car (T.T.) - $25.00–35.00
Reference: Left to right: NL28, NL 29, NL30

Manufacturer: Unknown (Japan)
Model: Friction power (set of 12)
Composition: Tin
Size: Each - 5¼"
Year: 1980s
Rarity: 1
Value: Set - $80.00–100.00; Individual - $20.00–25.00
Reference: NL31

Manufacturer: Taiyo (Japan)
Model: Bump 'n Go hot rod Snake and Mongoose, battery power
Composition: Tin, plastic
Size: Each - 10¼"
Year: 1970s
Rarity: 2
Value: Each - $75.00–100.00
Reference: Snake - NL32; Mongoose - NL32a

Manufacturer: Taiyo (Japan)
Model: Bump 'n Go w/blinking lights Corvette Stingray, Toronado, and Camaro Strip Blazer, battery operated
Composition: Tin
Size: Each - 7½"
Year: 1970s
Rarity: 2
Value: Each - $65.00–85.00
Reference: Left to right: NL33, NL34, NL34a

Manufacturer: Taiyo (Japan)
Year: 1980s
Model: Corvette Stingray, remote control
Rarity: 2
Composition: Plastic
Value: $35.00–45.00
Size: 8"
Reference: NL35

Manufacturer: Unknown **Model:** VW Van (right: lighter; left: coin box) **Composition:** Metal **Size:** 5" **Year:** 1980s **Rarity:** 3
Value: Each - $65.00–85.00 **Reference:** NL36, NL37

Manufacturer: Unknown (Japan)
Model: VW Bug, spring loaded
Composition: Tin
Size: 1½"
Year: 1980s
Rarity: 3
Value: $40.00–50.00
Reference: NL38

Manufacturer: Takara (Japan)
Model: Citroen van, spring loaded
Composition: Plastic
Size: 1½"
Year: 1980s
Rarity: 2
Value: $35.00–45.00
Reference: NL39TK

Manufacturer: Tomica Dandy (Japan)
Model: VW Van, #F23
Composition: Die-cast metal
Size: 1/43 scale
Year: 1980s
Rarity: 3
Value: $150.00–175.00
Reference: NL39

Manufacturer: Tomica (Japan)
Model: Citroen #F17
Composition: Die-cast metal
Size: 1/71 scale
Year: 1980s
Rarity: 2
Value: $60.00–80.00
Reference: NL40

Manufacturer: Tomica (Japan)
Model: Coca-Cola 100th Anniversary Isuzu #27
Composition: Die-cast metal
Size: 1/70 scale
Year: 1986
Rarity: 4
Value: $100.00–125.00
Reference: NL41

Manufacturer: Straco (Hong Kong)
Model: Wee People #9/735, friction power
Composition: Plastic
Size: 5½"
Year: 1977
Rarity: 1
Value: $40.00–50.00
Reference: NL42

Manufacturer: Unknown (Hong Kong) **Model:** Moving van truck, friction power **Composition:** Plastic **Size:** 5¼"
Year: 1970s **Rarity:** 4
Value: Coca-Cola only - $150.00–175.00; Set of 3 - $225.00–325.00 **Reference:** NL43a, NL43b, NL43c

Manufacturer: Petrel Toys (Hong Kong)
Model: Friction power
Composition: Plastic
Size: 3¾"
Year: 1970s
Rarity: 3
Value: $50.00–60.00
Reference: NL44

Manufacturer: Km (Hong Kong) **Model:** Cola truck, friction power **Composition:** Plastic
Accessories: 12 cases **Size:** 7¾" **Year:** 1980s **Rarity:** 3
Value: $150.00–250.00 **Reference:** NL45

Manufacturer: Parts Express (Hong Kong) **Model:** Tandem trailer **Composition:** Die-cast metal, plastic **Size:** 15"
Year: 1980s **Rarity:** 4 **Value:** Complete set - $175.00–275.00 **Reference:** NL46

Above:
Manufacturer: Durham Industries (Hong Kong)
Model: Zip-Along/Roll-Along trucks
Composition: Plastic **Size:** 4½"
Year: 1970s–80s **Rarity:** 1
Value: Zip-Along - $25.00–35.00; Roll-Along - $20.00–25.00
Reference: NL47, NL47a, NL47b
Note: Zip-Alongs have lime green hubs.

Right:
Manufacturer: Straco (Hong Kong)
Model: Soft drink truck **Composition:** Plastic
Size: 5" **Year:** 1980s
Rarity: 1 **Value:** $25.00–35.00
Reference: NL48

Manufacturer: Straco (Hong Kong)
Model: Fun Mates, wind-ups (2 variations)
Composition: Plastic
Size: 3¾"
Year: 1980s
Rarity: 1
Value: $25.00–35.00
Reference: NL49, NL49a

Manufacturer: Larami (Hong Kong)
Model: Cola Cars (2 variations)
Composition: Plastic
Size: 4"
Year: 1980s
Rarity: 1
Value: $25.00–35.00
Reference: Left to right: NL50, NL50a

Manufacturer: TFS (Hong Kong)
Model: Supermarket truck
Composition: Plastic
Size: 7½"
Year: 1970s
Rarity: 2+
Value: $20.00–25.00
Reference: NL51

Manufacturer: Clover Toys (Korea)
Model: #7602 (1 of 3 models)
Size: 5½"
Year: 1980s
Value: $40.00–50.00
Composition: Steel
Rarity: 3
Reference: NL52

Manufacturer: Kingstar (Korea)
Model: Ford Econoline E150
Composition: Die-cast metal
Size: 1/60 scale
Year: 1980s
Rarity: 2
Value: $35.00–45.00
Reference: NL53
Note: Also found in red, yellow, and silver.

Manufacturer: Miltan (Singapore)
Model: Delivery van
Composition: Die-cast metal
Size: 3¾"
Year: 1970s
Rarity: 4
Value: $175.00–225.00
Reference: NL54

Manufacturer: Maxwell Co. (India) **Model:** Delivery van **Composition:** Die-cast metal **Size:** 3¾" **Year:** 1970s
Rarity: 1 **Value:** $35.00–45.00 **Reference:** Left to right: NL55, NL55a

Manufacturer: Siku Eurobuilt (W. Germany) **Model:** #2918 Ford **Composition:** Die-cast metal **Accessories:** 8 yellow bins, hand truck
Size: 7½" (1⁄55 scale) **Year:** 1983 **Rarity:** 4 **Value:** $80.00–100.00 **Reference:** NL56
Note: This Siku was produced without authorization in extremely limited quantities and was subsequently replaced by the the licensed model shown below.

Manufacturer: Siku Eurobuilt (W. Germany) **Composition:** Die-cast metal **Year:** 1980s **Rarity:** 2 **Note:** Individual trucks listed below.

Model: Mack truck
Size: 12½" (1⁄55 scale)
Value: $35.00–45.00
Reference: NL57

Model: Ford cargo
Size: 7½" (1⁄55 scale)
Accessories: 12 white bins, hand truck
Value: $35.00–45.00
Reference: NL58

Model: Mercedes van
Size: 3½" (1⁄55 scale)
Accessories: 2 white bins
Value: 35.00–45.00
Reference: NL59

Model: VW mini van
Size: 3"
Value: $15.00–25.00
Reference: NL60VW

Model: VW pick-up
Size: 3"
Value: $15.00–25.00
Reference: NL60PU

Siku Old Timer pictured with actual c.1931 Coca-Cola delivery equipment identification guide.

Manufacturer: Siku Eurobuilt (W. Germany) **Model:** Old Timer "A" frame **Composition:** Die-cast metal **Accessories:** 12 cases
Size: 5¾" (1/55 scale) **Year:** 1980s **Rarity:** 1 **Value:** $50.00–65.00
Reference: NL61

Manufacturer: Europa/Lucky (Hong Kong)
Model: VW, motorized, battery operated
Composition: Plastic
Size: 4½"
Year: 1970s
Rarity: 3
Value: $125.00–150.00
Reference: NL62
Note: Produced for German market.

Manufacturer: Boyd (W. Germany)
Model: Flatbed trucks
Composition: Plastic
Accessories: Plastic case strip
Size: 1¾" (N scale)
Year: 1970s
Rarity: 3
Value: Each - $40.00–50.00
Reference: NL63
Note: Several variations shown.

Manufacturer: Herpa (W. Germany)
Composition: Plastic
Size: 1/87 scale
Year: 1980s
Rarity: 1
Value: $15.00–20.00
Reference: NL64

Manufacturer: Norev (France)
Composition: Die-cast metal
Size: 3¼"
Year: 1980s
Rarity: 1
Value: $10.00–15.00
Reference: NL65

Manufacturer: Roskopf (W. Germany)
Composition: Plastic
Size: 1/87 scale
Year: 1980s
Rarity: 1
Value: $15.00–20.00
Reference: NL66

Manufacturer: S.M.
Composition: Plastic, die-cast metal
Size: 8¼"
Year: 1980s
Rarity: 2
Value: $12.00–15.00
Reference: NL67

Manufacturer: Lemezarugyar (Hungary)
Model: Lendkerekes Mikrobusz, friction (a.k.a Combi-van)
Composition: Plastic
Size: 6¾"
Year: 1970s
Rarity: Blue and silver - 2; Red - 4
Value: Blue and silver - $60.00–80.00
Red - $80.00–100.00
Reference: NL68a, NL68b, NL68c

Manufacturer: Mucpo (Bulgaria)
Model: Delivery trucks (3 variations)
Composition: Plastic, steel
Accessories: 3 cases (center truck)
Size: 6"
Year: 1980s
Rarity: 4
Value: Each - $75.00–85.00
Reference: Left to right: NL69a, NL69b, NL69c
Note: These trucks are often believed to be Russian, but in fact are Bulgarian.

Manufacturer: Lion Car (Holland)
Size: 4¼"
Year: 1980s
Model: #49 Commer (group of 3 variations)
Rarity: 1
Value: $40.00–50.00
Composition: Die-cast metal
Reference: NL70

Manufacturer: Lion Car Blank w/Zaugg Modell Tag
Size: 4¼"
Year: 1980s
Model: Clown Models World Circus Edition
Rarity: 1
Value: $40.00–50.00
Composition: Die-cast metal
Reference: NL71

Right photo:

Manufacturer: Play Trucks (Greece) (top)
Model: #P1579
Composition: Plastic
Size: 11"
Year: 1980s
Rarity: 2
Value: $35.00–45.00
Reference: NL72

Manufacturer: Joy Toys (Greece) (bottom)
Model: #173
Composition: Plastic
Size: 8¾"
Year: 1980s
Rarity: 2
Value: $25.00–35.00
Reference: NL73

Center and left photos:

Manufacturer: Yaxon (Italy)
Model: #0300/3
Composition: Die-cast metal, plastic
Size: 13" (1/43 scale)
Year: 1970s
Rarity: 3
Value: $175.00–275.00
Reference: NL74

Manufacturer: R.E. El Toys (Italy) **Model:** Elettrico 4x4, battery operated **Composition:** Plastic **Size:** 9" **Year:** 1980s
Rarity: 3 **Value:** $80.00–100.00 **Reference:** NL75

Manufacturer: F. Godie, S.L. (Portugal)
Model: Bebidas candy
Composition: Plastic
Size: 3¾"
Year: 1970s
Rarity: 3 (With candy - 5)
Value: $100.00–150.00
Reference: NL76
Note: This toy truck was actually a candy container, see photos below. Rarely found with original candy.

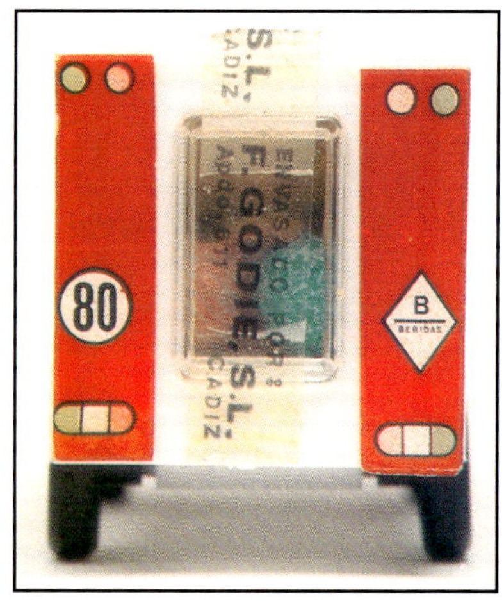

Manufacturer: Rico (Spain)
Model: Sanson Junior
Composition: Pressed steel
Size: 13½"
Year: 1970s
Rarity: 4
Value: $175.00–275.00
Reference: NL77

Manufacturer: Rico (Spain)
Model: Sanson/Bravo
Composition: Pressed steel
Size: 8½"
Year: 1970s
Rarity: 3
Value: $175.00–275.00
Reference: NL78

Manufacturer: Rico (Spain)
Model: Sanson/Bravo
Composition: Pressed steel
Size: 8½"
Year: 1970s
Rarity: 3
Value: $175.00–275.00
Reference: NL79

Reference: NL80, NL81

Manufacturer: Lesney (U.K.)
Model: Models of Yesteryear #Y12 (Left - #Y12 New York 75th Anniversary model)
Composition: Die-cast metal
Size: 3½"
Year: 1979
Rarity: 1; Anniversary - 4
Value: $50.00–60.00; Anniversary - $100.00–125.00
Reference: NL80, NL81

Reference: NL81

Manufacturer: Lesney/Matchbox (U.K.)
Model: #K31 Super Kings – Peterbilt trucks (3 variations)
Composition: Die-cast metal
Size: Each - 12¼"
Year: 1979
Rarity: 1
Value: $45.00–65.00
Reference: NL82

Manufacturer: Corgi (U.K.)
Model: #437 Chevy van
Composition: Die-cast metal
Size: 4¾"
Year: 1978
Rarity: 2
Value: $45.00–65.00
Reference: NL83

Right: Interior view of #437 Chevy van.

Manufacturer: Corgi (U.K.) **Model:** Juniors **Composition:** Die-cast metal **Size:** 2½–3" **Year:** 1970s **Rarity:** 1
Value: Each - $15.00–20.00 **Reference:** NL84, NL85, NL86, NL87
Note: Left to right: Chevy van, Cabover Leyland Terrier, Double decker Diamler Fleeting, Leyland Terrier Commemorative Edition

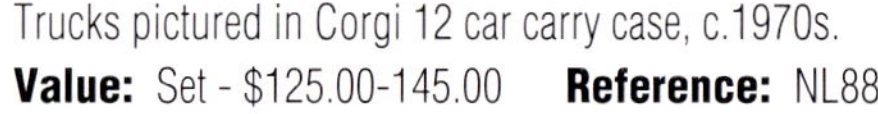
Trucks pictured in Corgi 12 car carry case, c.1970s.
Value: Set - $125.00-145.00 **Reference:** NL88

Manufacturer: Corgi (United Kingdom)
Size: Left to right: 3", 2¾", 3"
Value: Each - $10.00–15.00
Model: Juniors
Year: 1980s
Reference: NL89, NL90, NL91
Composition: Plastic, die-cast metal
Rarity: 1

Manufacturer: Yatming (China)
Model: Work trucks
Composition: Die-cast metal, plastic
Size: 2¾", trailer - 6¾"
Year: 1980s
Rarity: 1; trailer - 3
Value: $5.00–15.00, trailer - $15.00–20.00
Reference: NL92, NL92a

Manufacturer: Left - Larami (Hong Kong)
Composition: Plastic, die-cast metal
Year: 1980s
Value: Each - $8.00–20.00
Model: King of the Road
Size: 2¾"
Rarity: 1
Reference: NL93
Manufacturer: Right - UDC (Hong Kong)
Composition: Plastic, die-cast metal
Year: 1980s
Value: Each - $5.00–15.00
Model: Flash Truck
Size: 2¾"
Rarity: 1
Reference: NL93a

Note: Some colors are rarer than others which affects their value range.

Manufacturer: Unknown (Hong Kong)
Model: Super Wheels
Composition: Die-cast metal
Size: 2¾"
Year: 1980s
Rarity: 1
Value: $5.00–15.00
Reference: NL94
Note: Some colors are rarer than others which affects their value range

Manufacturer: Wheeler (Hong Kong)
Model: Super Trucks (set of 3)
Composition: Die-cast metal
Size: Each - 2¾"
Year: 1980s
Rarity: 1
Value: Set - $10.00–20.00
Reference: NL95

Manufacturer: Delhar Distributors (Hong Kong)
Composition: Die-cast metal
Year: 1980s
Value: Set - $15.00–20.00
Model: Super Wheels (set of 6)
Size: Each - 2¾"
Rarity: 1
Reference: NL96

Manufacturer: Ribbon Pencil (Japan)
Model: #330 Perfumed Eraser
Composition: Paper, plastic
Size: 2¾"
Year: 1980s
Rarity: 2
Value: $4.00–5.00
Reference: NL97

Manufacturer: WanDa (Taiwan)
Model: Container Set
Composition: Plastic
Size: 7½"
Year: 1980s
Rarity: 1
Value: $10.00–15.00
Reference: NL98

Left:
Manufacturer: Unknown (Hong Kong)
Model: Delivery vans
Composition: Plastic
Size: 2¾"
Year: 1970s
Rarity: 2
Value: $15.00–20.00
Reference: NL99

Right:
Manufacturer: Kosi Toy (Hong Kong)
Model: Truck set (set of 8)
Composition: Plastic
Size: Each - 3"
Year: 1980s
Rarity: 1
Value: $15.00–20.00
Reference: NL100

Left:
Manufacturer: Unknown (Hong Kong)
Model: Transport Fleet (set of 3)
Composition: Plastic, cardboard
Size: Each - 4½"
Year: 1980s
Rarity: 1
Value: Set - $25.00–35.00
Reference: NL101

Center:
Manufacturer: Happy Motors (Hong Kong)
Model: Container trucks (set of 3)
Composition: Plastic, cardboard
Size: Each - 4¾"
Year: 1980s
Rarity: 1
Value: Set - $25.00–35.00
Reference: NL102

Right:
Manufacturer: Elmar (Hong Kong)
Model: Heavy Haulers (set of 2)
Composition: Plastic, cardboard
Size: Each - 3¾"
Year: 1980s
Rarity: 1
Value: Set - $25.00–35.00
Reference: NL103

Blue cab truck w/antenna **Rarity:** 3 **Value:** $40.00–50.00 **Reference:** NL104

Manufacturer: Glasslite (Brazil)
Model: #5013.03 Expressinho
Composition: Plastic, tin
Size: 4⅛"
Year: 1980s
Rarity: 1
Value: $35.00–45.00
Reference: NL105

Manufacturer: Uirapuru (Brazil)
Model: Ford
Composition: Plastic
Accessories: 6 red cases w/clear bottles
Size: 7½"
Year: 1980s
Rarity: 2
Value: $35.00–45.00
Reference: NL106

Manufacturer: Brinquedos Rei (Brazil)
Model: #KS 54-C (Campeoes da Estrata)
Composition: Plastic, metal
Size: 7¼"
Year: 1980s
Rarity: 1
Value: $25.00–35.00
Reference: NL107

Manufacturer: Galgo (Argentina)
Model: Pro Truck (left - #30; right - #52)
Composition: Plastic, metal
Size: Left - 6¾"; right - 3¼"
Year: 1980s
Rarity: 1
Value: Left - $25.00–35.00; right - $20.00–25.00
Reference: NL109, NL110

Manufacturer: Impala (Mexico)
Composition: Plastic
Accessories: 8 cases w/ black bottles
Size: 7"
Year: 1980s
Rarity: 2
Value: $15.00–25.00
Reference: NL111

Manufacturer: Plasticos Delta S.A. (Mexico)
Composition: Plastic
Accessories: 8 cases w/ multicolored bottles
Size: 7"
Year: 1980s
Rarity: 2
Value: $15.00–25.00
Reference: NL112

Manufacturer: Beka (Mexico)
Composition: Plastic
Accessories: 8 cases w/red, green, yellow bottles
Size: 7"
Year: 1980s
Rarity: 3
Value: $25.00–35.00
Reference: NL113

Manufacturer: Impala (Mexico)
Composition: Plastic
Accessories: 8 cases w/ black bottles
Size: 7"
Year: 1980s
Rarity: 2
Value: $15.00–25.00
Reference: NL114

Manufacturer: Impala (Mexico)
Composition: Plastic
Accessories: 8 cases w/black bottles
Size: 7"
Year: 1980s
Rarity: 2
Value: Each - $15.00–25.00
Reference: NL115

Manufacturer: Unknown (Mexico) **Model:** Unknown **Composition:** Plastic **Accessories:** 2 color cases **Size:** 5¼"
Year: 1980s **Rarity:** 2 **Value:** $12.00–15.00 **Reference:** NL116

Manufacturer: Impala (Mexico) **Model:** Delivery trucks **Composition:** Plastic **Accessories:** 12 cases
Size: 11" **Year:** 1980s **Rarity:** 2 **Value:** $25.00–35.00
Reference: NL117

Manufacturer: Uni-Plast (Mexico) **Model:** Futuristic **Composition:** Plastic **Accessories:** 12 cases
Size: 11" **Year:** 1980s **Rarity:** 3 **Value:** $75.00–85.00
Reference: NL118

Manufacturer: Uni-Plast (Mexico)
Model: GMC pickup
Composition: Plastic
Size: 12½"
Year: 1980s
Rarity: 3
Value: $65.00–85.00
Reference: NL119

Manufacturer: Uni-Plast (Mexico)
Model: #302 Repartidora Vanet
Composition: Plastic
Size: 9½"
Year: 1978–79
Rarity: 2
Value: $65.00–75.00
Reference: NL120

Manufacturer: Tente (Mexico)
Model: Lego-type, 2 variations
Composition: Plastic
Size: 2¾"
Year: 1978–79
Rarity: 3
Value: Each - $40.00–50.00 (complete)
Reference: NL121

Manufacturer: Uni-Plast, (Mexico)
Model: GMC
Composition: Plastic
Accessories: 8 yellow or white cases
Size: 7½"
Year: 1970s
Rarity: 4
Value: $275.00–325.00
Reference: NL122

Manufacturer: Uni-Plast (Mexico)
Model: #243, Repartidora
Composition: Plastic
Size: 4"
Year: 1970s
Rarity: 3
Value: $150.00–175.00
Reference: NL123

Manufacturer: Lemy (Mexico)
Model: #323, VW pick-up, friction power
Composition: Tin
Accessories: 4 red cases w/black bottles
Size: 9"
Year: 1970s
Rarity: 3
Value: $150.00–250.00
Reference: NL124
Note: Lemy purchased the original dies from the West German firm, Tipp & Co., to produce the #323 Coca-Cola VW pick-up in the mid-1970s. Interestingly, they both shared the same model number 323. See page 83.

Manufacturer: Mattel (USA)
Rarity: 3-5
Model: Hot Wheels Mongoose and The Snake II
Value: Each - $100.00–200.00
Composition: Die-cast metal
Reference: NL125, NL125a
Size: 3"
Year: 1970s

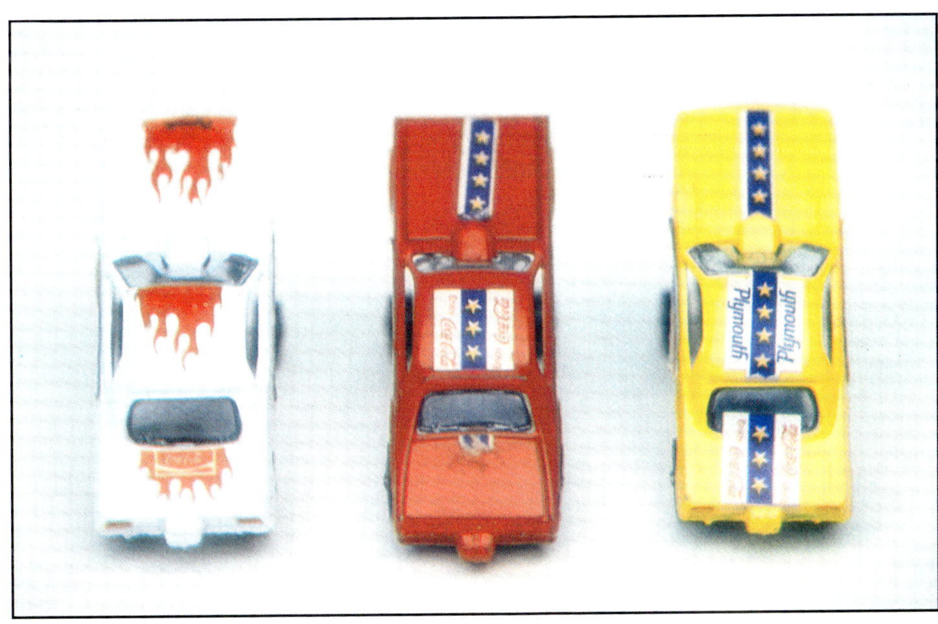

Mattel Hot Wheels with Coca-Cola logos appeared on two different models and several variations of the Mongoose and the Snake, which came in four color versions – red, white, blue, and yellow. Some variations are much rarer than others. Hot Wheels are becoming increasingly collectible! **Reference:** NL125

Manufacturer: Testors (USA)
Model: Sprite Special, fuel engine (2 variations)
Composition: Plastic
Size: 12"
Year: 1970s
Rarity: 2
Value: Each - $80.00–100.00
Reference: NL126, NL126a

Reference: NL126

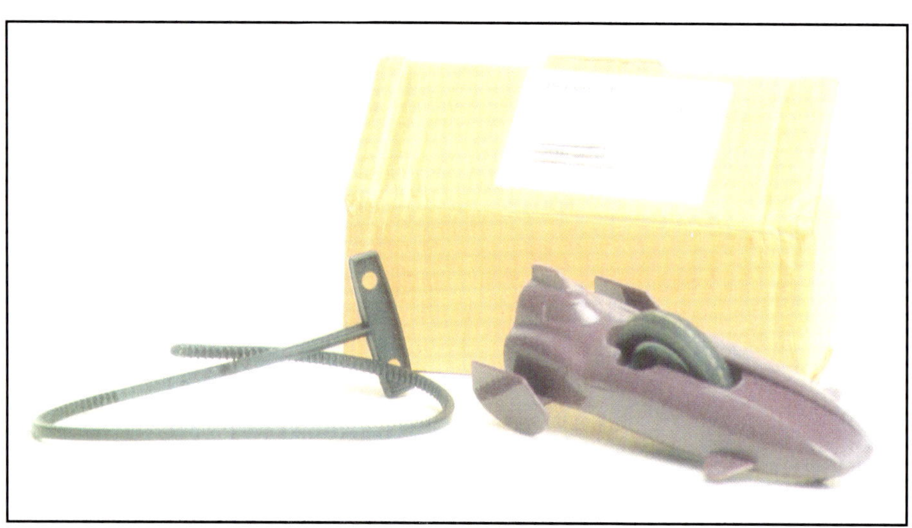

Reference: NL126a

Manufacturer: Kenner (USA)
Model: Hi-C Brand Mini SSP Racer (Super Sonic Power)
Composition: Plastic
Size: 6½"
Year: 1970
Rarity: 4
Value: $60.00–80.00
Reference: NL127

Manufacturer: Winross (USA) **Model:** White tractor trailer **Composition:** Die-cast metal **Size:** 9⅜" **Year:** 1970s **Rarity:** 3
Value: $150.00–175.00 **Reference:** NL128 **Note:** This model came in 3 rear door variations – red, silver, and white (not shown).

Manufacturer: Nylint (USA) **Model:** Top: #910 Cadet 18-wheeler; bottom: #911Z GMC tanker **Composition:** Pressed steel
Size: 21" **Year:** 1980s **Rarity:** 2 **Value:** Each - $50.00–60.00 **Reference:** NL129, NL130

This is a small sampling of the cardboard trucks that appeared on the market in the 1980s. Most of these share the same $8.00–12.00 value.

Manufacturer: Zippy's (USA)
Size: 8"
Rarity: 1
Reference: NL131

Manufacturer: Unknown
Size: 13"
Rarity: 1
Reference: NL132

Manufacturer: CCCI
Size: 18"
Rarity: 1
Reference: NL133

Manufacturer: Sweetcentre
Size: 7"
Rarity: 1
Reference: NL134

Manufacturer: Toys For Adults (TFA) (USA) **Model:** Cushions (Domino's Pizza Indy, Bobby Allison NASCAR, Bobby Allison Signature NASCAR)
Composition: Cloth **Size:** 15–20" **Year:** Late 1970–1980s **Rarity:** 3 **Value:** NASCAR - $100.00–125.00 each; Indy - $80.00–100.0
Reference: NL135, NL136, NL137

Manufacturer: Huron Product Co. (USA)
Model: Bank
Composition: Plastic
Size: 6½"
Year: 1970s
Rarity: 1
Value: $15.00–18.00
Reference: NL138

Manufacturer: Unknown
Model: VW van
Composition: Plastic
Size: 6½"
Year: 1980s
Rarity: 2
Value: $15.00–18.00
Reference: NL139

Manufacturer: Ertl (USA) **Model:** Cale Yarborough Hardee NASCAR **Composition:** Plastic **Size:** 12¼" **Year:** 1980s
Rarity: 1 **Value:** $15.00–20.00 **Reference:** NL140
Note: These cars also appeared in sets w/4 x 4s and trailers.

Reference: NL141

Reference: NL142

Above Left:
Manufacturer: Unknown (Hong Kong)
Composition: Plastic
Year: 1980s
Value: $25.00–35.00
Model: Diet Coke, friction power
Size: 3¾"
Rarity: 3
Reference: NL141

Above Right:
Manufacturer: Playart (Hong Kong)
Composition: Die-cast metal
Year: 1980s
Value: $80.00–100.00
Model: Diet Coke, futuristic
Size: 2½"
Rarity: 5
Reference: NL142

Note: This vehicle is 1 of 250 produced for the Diet Coke World Premiere Festivities Dinner in New York City, July 29, 1982.

Reference: NL143, NL144

Right:
Manufacturer: Unknown (Hong Kong)
Composition: Metal
Year: 1980s
Value: $15.00–25.00
Model: Quartz digital clock
Size: Top - 6⅝"; bottom - 4½"
Rarity: 1
Reference: NL143, NL144

Manufacturer: Van Goodies (Canada)
Model: Denimachine
Composition: Faux wood
Size: 12"
Year: Late 1970s
Rarity: 4
Value: $150.00–250.00
Reference: NL145

Manufacturer: Toystalgia, Inc. (USA) **Model:** Bank
Composition: Wood **Size:** 7¼" **Year:** 1980s
Rarity: 1 **Value:** $20.00–25.00
Reference: NL146

Manufacturer: Toystalgia, Inc. (USA) **Model:** Radio
Composition: Wood, plastic **Size:** 7¼" **Year:** 1979 **Rarity:** 1
Value: $25.00–35.00 **Reference:** NL147

Manufacturer: Unknown **Model:** 100th Anniversary truck **Composition:** Laser etched wood **Size:** 20"
Year: 1986 **Rarity:** 4 **Value:** $450.00–650.00 **Reference:** NL148
Note: This 100th Anniversary truck was produced in *extremely* limited quantities.

Monster Truck
MADNESS

Manufacturer: Lapin (USA) **Model:** *"Competitor Crusher"* Super Van **Composition:** Plastic **Size:** 18" **Year:** 1980s
Rarity: 2 **Value:** $65.00–85.00 **Reference:** NL149

COLLECTORS' EDITIONS,
RACING CARS,
MODEL KITS

COLLECTORS' EDITION, RACING CARS, MODEL KITS

"Collectors' edition" is the term used for vehicles produced in very limited quantities, usually no more than 2,000, and more often, only 500 of each model, exclusively made for the collectors' market. All too often, however, these are not licensed or authorized by the owner of the trademark they display. The process starts with a blank acquired from a major manufacturer, (i.e) Lledo, Matchbox, Solido, Ralstoy, Corgi, etc. The blank is then painted to specifications and decaled or silk-screened accordingly. It is not unusual to see a single model displaying the advertising of 25 to 30 different companies. An example of this is precisely what happened with the first five Lledos that were issued bearing the Coca-Cola trademark.

In 1983, approximately 1,440 blanks were purchased from Lledo and shipped to the United States, where they were silk-screened with "Coca-Cola, At Soda Fountains." These trucks were then reinserted into their original Lledo boxes and sold to collectors with no mention that they had been produced on the secondary market. Lledo, displeased with the quality of the silk-screening, offered to produce a better quality model, more up to their standards, provided the company obtained written approval from the Coca-Cola Company for the use of their trademark. This company somehow produced the "proper" authorization, and Lledo subsequently manufactured a new Coca-Cola model exclusively for them. Several months later, Lledo produced three additional models, again exclusively for this company that claimed to have world rights. Interestingly, The Coca-Cola Company had no record of anyone other than Siku having the rights to their trademark on die-cast Coca-Cola toy trucks in the United States. What ensued is unclear, but shortly thereafter a Florida corporation, Hartoy, Inc., became the exclusive licensed distributor for Coca-Cola brand die-cast vehicles. The first five models produced without proper authorization and before Hartoy are the ones we feel will become the most valuable Lledo Coca-Cola models.

This chapter includes some of these more famous, and infamous, collectors edition vehicles that we have come to know over the past 15 years or so. Miniature Toys Incorporated (a.k.a. Nostalgic Miniatures) is probably responsible for some of the more interesting ones. Most Coca-Cola collectors are very familiar with these little yellow 1/43 scale die-casts produced by MTI from the mid-1970s to the mid-1980s. MTI used blanks or models by M.V.M., Eligor, Solido, Maxitoy, Brooklin Models, and Siku, as well as others, to produce the many vehicles in their Coke repertoire. In the mid-1980s, amongst threats of

Manufacturer: M.I.C. (cab), D.S. (body) (USA) **Model:** B-Mack **Composition:** Cast aluminum **Accessories:** 12 cases
Size: 18½" **Year:** 1980s **Rarity:** 4 **Value:** $650.00–850.00
Reference: CD1

lawsuits for the unauthorized use of the Coca-Cola trademark, MTI ceased the production of their line of miniature Coca-Cola vehicles. According to sources at the Coca-Cola Company, MTI was given a time and date by which they had to liquidate their inventory and could no longer use the trademark. The MTI vehicles are perhaps the handsomest and most desirable of all these collectors editions. A limited number of pewter nostalgic models was also produced.

Other famous models from Italy, France, and Germany are pictured in this section, as well as a comprehensive selection of the enigmatic Smith trucks from Great Britain and a few other models of unknown origin. The French vehicles pictured are just a small sampling of the vast quantity of such vehicles produced in France in the early 1980s. They serve as a reminder of just how easily a vehicle can be produced. These esoteric models were intended exclusively to satisfy the Coca-Cola collectors' market and were gobbled up by hungry collectors as quickly as they appeared at shows. It seems that all you need is a little imagination and an ability to recognize a niche in the market to create a Coca-Cola vehicle.

The race car section includes several very limited production "trans-kits," including the BMW 3000, the Alpine Rallye 1600, and the Porsche 935 by Solido, and models by Starter, Mini Racing, Record, Faster, and Burago. Lastly, we have also included the more famous model kits in this chapter. There are many additional model kits available than we were able to picture in this section and most are stock cars and/or race cars.

Manufacturer: Unknown (USA) **Year:** 1980s **Model:** Commer van w/Coke machine **Rarity:** 3 **Composition:** Die-cast metal **Value:** Each - $30.00–40.00 **Size:** N scale **Reference:** CD2

Manufacturer: Brekina (W. Germany) **Rarity:** 3 **Model:** #DKWF7 **Value:** $35.00–45.00 **Composition:** Plastic **Reference:** CD3 **Size:** N scale **Year:** 1980s

Manufacturer: Nostalgic Miniature (USA) **Model:** 1930s Ford Van **Composition:** Die-cast metal **Size:** N scale
Year: 1980s **Rarity:** 4 **Value:** $80.00–100.00 **Reference:** CD4

Manufacturer: Maxitoys (Holland) **Model:** Oldtimer Models (Ford) **Composition:** Metal **Size:** 11"
Year: 1980s **Rarity:** 4 (500 made) **Value:** $425.00–525.00 **Reference:** CD5

Manufacturer: Nostalgic Miniature (USA) **Composition:** Die-cast metal **Rarity:** 3 **Value:** $100.00–125.00

Model: 1931 Ford AA
Year: 1978
Size: 5¼"
Reference: CD6

Model: 1931 Ford AA
Year: 1978
Size: 5¼"
Reference: CD7

Model: 1931 Ford AA
Year: 1978
Size: 5¼"
Reference: CD8

Model: 1936 Ford Panel
Year: 1981
Size: 4¼"
Reference: CD9

Model: 1936 Ford Panel
Year: 1981
Size: 4¼"
Reference: CD10

Model: 1936 Dodge
Year: 1983
Size: 5"
Reference: CD11

Manufacturer: Nostalgic Miniature (USA) **Model:** 1931 Ford Model A **Composition:** Die-cast metal **Year:** Marked 1974, distributed in 1976

Model: Panel truck
Size: 3¼"
Rarity: 3
Value: $100.00–120.00
Reference: CD12

Model: Embossed logo
Size: 3¼"
Rarity: 4
Value: $125.00–145.00
Reference: CD13

Model: Pick-up
Size: 3⅜"
Rarity: 3
Value: $100.00–120.00
Reference: CD14

Model: Coupe
Size: 3⅛"
Rarity: 3
Value: $100.00–120.00
Reference: CD15

Bottom row, left to right:
Manufacturer: Mini Marque (England)
Model: #43, 1936 Ford V8
Composition: Die-cast metal
Size: 3¼"
Year: 1980s
Rarity: 3
Value: $100.00–125.00
Reference: CD16

Manufacturer: MVC (USA)
Model: 1937 Studebaker coupe
Composition: Die-cast metal
Size: 3¾"
Year: 1980s
Rarity: 3
Value: $85.00–100.00
Reference: CD17

Manufacturer: MVC (USA)
Model: 1937 Studebaker coupe
Composition: Die-cast metal
Size: 3¾"
Year: 1980s
Rarity: 3
Value: $85.00–100.00
Reference: CD18

Manufacturer: P.A.C. (USA)
Model: Chevrolet
Composition: Die-cast metal
Size: 3⅜"
Year: 1980s
Rarity: 2
Value: $45.00–65.00
Reference: CD19

Manufacturer: Nostalgic Miniatures (USA)
Model: Federal truck
Composition: Die-cast metal
Size: 2¾"
Year: 1977
Rarity: 3
Value: $60.00–80.00
Reference: CD20

Original Coca-Cola artwork of the type of vehicles often depicted by MTI.

Manufacturer: Nostalgic Miniatures (USA)
Model: 1954 International
Composition: Die-cast metal
Size: 4"
Year: 1982
Rarity: 3
Value: $100.00–125.00
Reference: CD21
Note: One of two variations produced in yellow. There is also a red and white version. **Rarity:** 5 **Value:** $150.00–200.00.

Manufacturer: Nostalgic Miniature (USA) **Model:** Mack truck **Composition:** Die-cast metal **Size:** 4¾"
Year: 1983 **Rarity:** 3 **Value:** $80.00–100.00 **Reference:** CD22

Manufacturer: Nostalgic Miniature (USA)
Model: Limited edition plaques (500 each)
Composition: Wood, metal
Size: 4½"
Year: 1980s
Rarity: 4
Value: $250.00–350.00
Reference: CD23

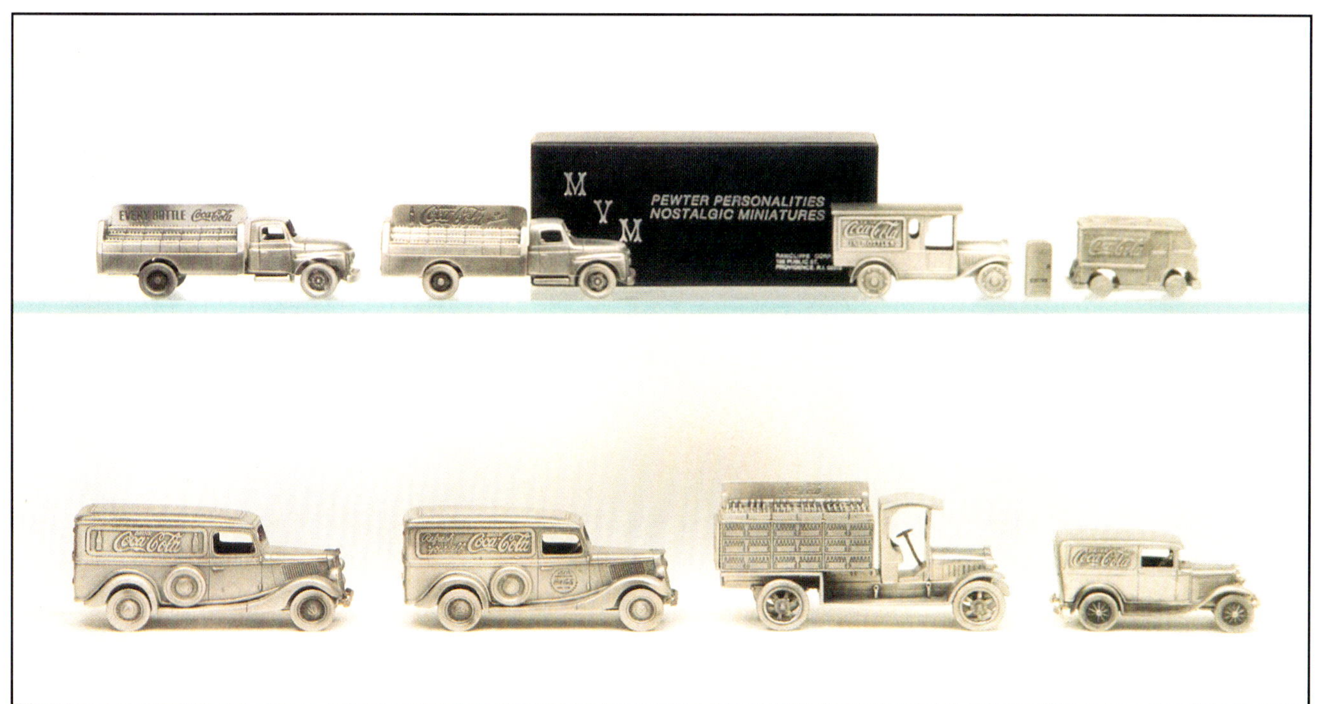

Pewter Nostalgic Miniatures made by MVM. Individual listings are below.

Model: #0123 1954 International
Size: 3⅞"
Year: 1980s
Rarity: 3
Value: $125.00–150.00
Reference: CD24

Model: #0123 1954 International
Size: 3⅞"
Year: 1980s
Rarity: 3
Value: $125.00–150.00
Reference: CD25

Model: #0717 Federal truck
Size: 2¾"
Year: 1980s
Rarity: 3
Value: $80.00–100.00
Reference: CD26

Model: #5 Ramcliffe
Size: 2"
Year: 1980s
Rarity: 2
Value: $35.00–45.00
Reference: CD27

Model: #0171 1936 Ford Panel
Size: 4¼"
Year: 1980s
Rarity: 3
Value: $125.00–150.00
Reference: CD28

Model: #0178 1936 Ford Panel 50th Anniversary
Size: 4¼"
Year: 1980s
Rarity: 3
Value: $150.00–175.00
Reference: CD29

Model: #007 (available only on plaque)
Size: 4½"
Year: 1980s
Rarity: 4
Value: $150.00–175.00
Reference: CD30

Model: 1931 Ford Model A
Size: 3¼"
Year: 1970s
Rarity: 3
Value: $125.00–150.00
Reference: CD31

Detail photo of #0178 1936 Ford Panel- 50th Anniversary.
Reference: CD29

Manufacturer: Lledo (U.K.) **Model:** Days Gone By **Composition:** Die-cast metal **Year:** 1980s

Model: #DG6-DG8 (silver box)
Size: 2½"
Rarity: 4
Value: $75.00–85.00
Reference: CD32
Note: 1,440 produced

Model: #DG6-DG8
Size: 2½"
Rarity: 2
Value: $20.00–25.00
Reference: CD33

Model: #DG7-DG9
Size: 3"
Rarity: 2
Value: $20.00–30.00
Reference: CD34

Model: #DG6-DG8
Size: 3"
Rarity: 1
Value: $20.00–25.00
Reference: CD35

Model: #DG2-DG3
Size: 3"
Rarity: 1
Value: $20.00–25.00
Reference: CD36

Model: #DG6-DG8
Size: 3"
Rarity: 1
Value: $20.00–25.00
Reference: CD37

(These three vehicles, CD35-37, as a set – **Rarity:** 2 **Value:** $75.00–80.00)

Model: #DG7-DG9
Size: 3"
Rarity: 2
Value: $20.00–25.00
Reference: CD38

Model: #DG7-9-13-14
Size: 3"
Rarity: 1
Value: $12.00–15.00
Reference: CD39

Model: #DG7-9-13-14
Size: 3"
Rarity: 1
Value: $12.00–15.00
Reference: CD40

Manufacturer: Lledo (U.K.) **Model:** Days Gone By **Composition:** Die-cast metal **Year:** 1980s **Rarity:** 1 **Value:** $10.00–15.00

Model: #DG15
Size: 3⅜"
Reference: CD41

Model: #DG16
Size: 3¼"
Reference: CD42

Model: #DG11
Size: 4½"
Reference: CD43

Model: #DG6-DG8
Size: 2½"
Reference: CD44

Model: #DG21
Size: 3"
Reference: CD45

Model: #DG20
Size: 3⅝"
Reference: CD46

Model: #DG6-DG8
Size: 2½"
Reference: CD47

Manufacturer: Eligor (France)
Model: 1932 Fords
Composition: Die-cast metal
Size: 1/43 scale
Year: 1980s
Rarity: 3
Value: Each - $65.00–85.00; Set - $125.00–150.00
Reference: CD47, CD47a

Manufacturer: Brooklin Models (U.K.)
Model: 1936 Dodge van #16, black (on box); 1936 Dodge van #16, yellow; 1940 Ford delivery sedan #9
Composition: Die-cast metal
Size: 1/43 scale
Year: 1980s
Rarity: Yellow - 3; Black - 5
Value: Yellow - $125.00–150.00; Black - $175.00–200.00
Reference: Left to right: CD48, CD49, CD49a

Manufacturer: MTI (top); Kennedy (bottom)
Model: Tootsietoys copy
Composition: Die-cast metal
Size: 3 1/8"
Year: 1980s
Rarity: 2
Value: Top - $35.00–45.00; bottom - $30.00–40.00
Reference: CD50 (top), CD50a (bottom)

The die-cast metal trucks shown were produced in West Germany in *extremely* limited quantity (50 per model).

Manufacturer: Top row: Solido, Matchbox, Solido; Bottom row: Matchbox, Corgi (2), Matchbox
Composition: Die-cast metal **Size:** 1⁄43 scale **Year:** 1980s **Rarity:** 5 **Value:** Each - $150.00–175.00
Reference: Top row (left to right) CD51, CD52, CD53; Bottom row (left to right): CD54, CD55, CD56, CD57

Manufacturer: Matchbox (U.K.) **Model:** Models of Yesteryear, right: Crossley Y-13; left: Talbot Y-5 **Composition:** Die-cast metal
Size: 1⁄43 scale **Year:** 1980s **Rarity:** 4 **Value:** Each - $80.00–100.00 **Reference:** CD58, CD59
Note: These models were produced for the European market.

Manufacturer: DGM (England) **Model:** Morris trucks **Composition:** Die-cast metal **Size:** 1/43 scale **Year:** 1980s
Rarity: 3 **Value:** Each - $60.00–75.00 **Reference:** Top row: (left to right) CD60, CD60a, CD60b, CD60c
Bottom row: (left to right) CD60d, CD60e, CD60f, CD60g
Note: Color, body style, and wheel variations pictured.

Manufacturer: DGM (England) **Model:** Morris trucks **Composition:** Die-cast metal **Size:** 1/43 scale **Year:** 1980s
Rarity: 3 **Value:** Each - $50.00–75.00 **Reference:** CD61, CD61a, CD61b **Note:** Wheel variations pictured.

Manufacturer: Zaugg (Germany)
Model: #13 - '51 Chevy station wagon (very limited)
Composition: Die-cast metal
Size: 1/43 scale
Year: 1980s
Rarity: 4
Value: $140.00–160.00
Reference: CD62

Manufacturer: Autohobby (Italy)
Model: Citroen
Composition: Die-cast metal
Size: 1/43 scale
Year: 1980s
Rarity: 3–4
Value: Each - $100.00–125.00
Reference: CD63, CD64

Manufacturer: Auto Buff (U.K.)
Model: Collectors' Edition '53 Ford
Composition: Die-cast metal
Size: 1/43 scale
Year: 1980s
Rarity: 3
Value: $100.00–125.00
Reference: CD65

Manufacturer: Micromodels (Australia)
Model: F.J. Holden van
Composition: Plastic
Size: 1/43 scale
Year: 1970s
Rarity: 4
Value: $225.00–325.00
Reference: CD66

Manufacturer: Traxx (Australia)
Model: F.J. Holden van
Composition: Plastic
Size: 1/43 scale
Year: 1980s
Rarity: 4
Value: $175.00–275.00
Reference: CD67

Manufacturer: So. Australia Coca-Cola Bottlers
Model: Commemorative truck/bottle
Composition: Cardboard
Size: 8"
Year: 1980s
Rarity: 3
Value: $50.00–75.00
Reference: CD68

Manufacturer: STL (France) **Model:** Peugeot, left: #203; right: #403 **Composition:** Resin composite **Size:** 1/43 scale
Year: 1970s **Rarity:** 5 **Value:** Set - $325.00–375.00 **Reference:** CD69

Manufacturer: Solido (France) **Model:** Renault **Compositlon:** Die-cast metal **Size:** 1/43 scale **Year:** 1980s
Rarity: 1-2 **Value:** Black box - $50.00–75.00; others - $35.00–45.00 **Reference:** CD70

Manufacturer: Solido (France) **Model:** French Collectors' Edition, left: Cadillac; right: Buick
Composition: Die-cast metal **Size:** 1/43 scale **Year:** 1980s **Rarity:** 3–4 **Value:** Each - $80.00–100.00
Reference: CD71, CD72

Manufacturer: Solido (produced for French market)
Model: Left - '50 Chevy; right - Chrysler Windsor
Composition: Die-cast metal
Size: 1/43 scale
Year: 1980s
Rarity: 3
Value: Each - $50.00–60.00
Reference: CD73, CD73a

Note: There have been many recently mass-produced Solido 1/43 scale Cola-Cola models similar to the above. These models appear in a red Coca-Cola box.
Rarity: 1 **Value:** $15.00–20.00.

Manufacturer: Solido (France)
Model: #32 Citroen
Composition: Die-cast metal
Size: 1/43 scale
Year: 1980s
Rarity: 2
Value: $65.00–75.00
Reference: CD74

Manufacturer: Solido (France)
Model: Collector Models
Composition: Die-cast metal
Size: 1/43 scale
Year: 1980s
Rarity: 3
Value: Each - $50.00–75.00
Reference: CD75, CD75a, CD75b, CD75c, CD75d

Manufacturer: Solido (France)
Model: Italian Collectors - #4423
Composition: Die-cast metal
Size: 1/43 scale
Year: 1980s
Rarity: 3
Value: $65.00–85.00
Reference: CD76

Manufacturer: Solido (France)
Model: Italian Collectors - #4413
Composition: Die-cast metal
Size: 1/43 scale
Year: 1980s
Rarity: 3
Value: $65.00–85.00
Reference: CD77

Manufacturer: Brumm-Glamour (Italy) **Model:** Fiat **Composition:** Die-cast metal **Size:** 1/43 scale
Year: 1980s **Rarity:** 3 **Value:** Set - $100.00–125.00; Each - $50.00–65.00
Reference: CD78

Manufacturer: Old Cars/Glamour (Italy) **Model:** Iveco, Fiat **Composition:** Die-cast metal **Size:** 1/43 scale
Year: 1980s **Rarity:** 3 **Value:** Each - $60.00–80.00 **Reference:** CD79, CD79a
Note: Two variations shown. Left -Iveco; right - Fiat.

Manufacturer: Budgie (U.K.) **Model:** VW pick-up **Composition:** Die-cast metal **Size:** 1/43 scale **Year:** 1980s
Rarity: 1 **Value:** $30.00–40.00 **Reference:** CD80

Manufacturer: Ralstoy (USA)
Model: #22
Composition: Die-cast metal
Size: 4½"
Year: 1980s
Rarity: 3–4
Value: $100.00–125.00
Reference: CD81
Note: 250 made

Manufacturer: Minibrindes (Brazil)
Model: Mini Mack
Composition: Metal
Size: 7"
Year: 1980s
Rarity: 3
Value: $60.00–75.00
Reference: CD82

Manufacturer: MTI (possibly)
Model: Mack truck
Composition: Die-cast metal, plastic
Size: 6⅝"
Year: 1980s
Rarity: 3
Value: $60.00–75.00
Reference: CD83

The following selection of die-cast trucks were manufactured by Smith in the United Kingdom. These exquisitely detailed Coca-Cola tractor trailers were all handcrafted in the mid-1980s. Because of their very limited production and rarity they are being rated at 4 on our scale and valued at $150.00–250.00 each. Smith trucks are extremely fragile.

Smith trucks are handsome and quite detailed.

Three variations of Smith Mack trucks. **Reference:** CD84, CD85, CD86

Five variations of Smith cab-over. **Reference:** CD87, CD88, CD89, CD90, CD91

Close-up of Smith cab-over truck.

Two variations of Smith cab-over. **Reference:** CD92, CD93

Three variations of Smith cab-over. **Reference:** CD94, CD95, CD95a

Racing Cars

Manufacturer: Starter (France)
Year: 1980s
Model: Hardee's Ford Thunderbird
Rarity: 3
Composition: Resin composite
Value: $60.00–80.00
Size: 1/43 scale
Reference: CD96

Top Row:
Manufacturer: Tomica Dandy (TOMY)
Year: 1980s
Model: Nissan (3 variations)
Rarity: 1
Composition: Die-cast metal
Value: Each - $35.00–45.00
Size: 1/43 scale
Reference: CD97, CD98, CD99

Bottom Row:

Manufacturer: Vitesse (Portugal)
Model: Porsche 956
Composition: Die-cast metal
Size: 1/43 scale
Year: 1980s
Rarity: 3
Value: $50.00–60.00
Reference: CD100

Manufacturer: Corgi (U.K.)
Model: Porsche 956
Composition: Die-cast metal
Size: 1/43 scale
Year: 1980s
Rarity: 1
Value: $25.00–35.00
Reference: CD101

Manufacturer: Starter (France)
Model: Porsche 956
Composition: Resin composite
Size: 1/43 scale
Year: 1980s
Rarity: 3
Value: $60.00–80.00
Reference: CD102

Manufacturer: Mini Racing Model (France)
Model: Porsche 956 (Car #76 - #0098; Car #5 - #0094)
Composition: Die-cast metal
Size: 1/43 scale
Year: 1980s
Rarity: 3
Value: $40.00–60.00
Reference: CD103, CD104

Manufacturer: Right - Starter; Left - Faster
(Both from France)
Model: Right - Porsche 917; Left - Porsche 962
Composition: Resin composite
Size: 1/43 scale
Year: 1980s
Rarity: 3
Value: Starter - $60.00–80.00; Faster - $50.00–60.00
Reference: CD105, CD106

Manufacturer: Left - Alezan; Left - Record Model
(Both from France)
Model: Left - Nissan; Right - Porsche 935
Composition: Resin composite
Size: 1/43 scale
Year: 1980s
Rarity: 3
Value: Each - $40.00–60.00
Reference: CD107, CD108

Manufacturer: On box - Burago (Italy)
Model: Porsche 935
Composition: Metal
Size: 1/43 scale
Year: 1980s
Rarity: 2
Value: $60.00–80.00
Reference: CD109

Manufacturer: Foreground - Solido trans-kit (France)
Model: Porsche 935 #1032
Composition: Die-cast metal
Size: 1/43 scale
Year: 1980s
Rarity: 4
Value: $125.00–150.00
Reference: CD110

Manufacturer: Solido trans-kit (France)
Model: Alpine Rallye 1600 #181
Composition: Die-cast metal
Size: 1/43 scale
Year: 1980s
Rarity: 4
Value: $125.00–150.00
Reference: CD111

Manufacturer: Solido trans-kit (France)
Model: BMW 3000 #75
Composition: Die-cast metal
Size: 1/43 scale
Year: 1980s
Rarity: 4
Value: $200.00–250.00
Reference: CD112
Note: Only 50 made with trans-kit

Model Kits

Manufacturer: Model Products Corp. (USA)
Model: Vending Machine
Composition: Plastic
Size: 4½"
Year: 1970s
Rarity: 4
Value: $100.00–120.00
Reference: Mk1

Completed model shown with box and instructions.

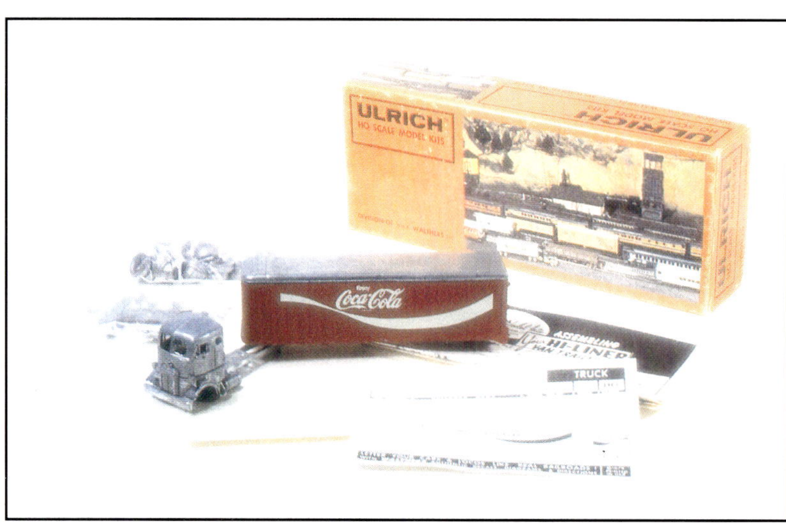

Manufacturer: Ulrich (USA)
Model: Cab-over
Composition: Die-cast kit
Size: 7"
Year: 1970s
Rarity: 3
Value: $65.00–85.00
Reference: Mk2

Manufacturer: Entex (USA)
Model: 1913 Model T Van
Composition: Wood, plastic
Size: 1/16 scale
Year: 1970s
Rarity: 2
Value: MIB - $60.00–75.00
Reference: Mk3

Manufacturer: Lasertec (USA)
Model: 100th Anniversary Collector's Edition
Composition: Cardboard
Size: 16¾"
Year: 1980s
Rarity: 3
Value: $25.00–35.00
Reference: Mk4

Manufacturer: Unknown (Japan)
Model: Fold-up truck
Composition: Cardboard
Size: 5" x 30"
Year: 1980s
Rarity: 4
Value: $60.00–80.00
Reference: Mk5

Top:
Manufacturer: AMT (USA)
Model: Joie Chitwood Thrill Show Camaro #T441
Rarity: 2 **Value:** $40.00–60.00 **Reference:** Mk6

Middle:
Manufacturer: AMT (USA)
Model: Cale Yarborough Hardees Monte Carlo
Rarity: 2 **Value:** $40.00–60.00 **Reference:** Mk7

Manufacturer: Revell (USA)
Model: Henry J Gas Coupe – Dyno-mite "J"
Rarity: 2 **Value:** $40.00–60.00 **Reference:** Mk8

Bottom:
Manufacturer: Craft Master (USA)
Model: Don Nicholson's Eliminator Funny Car
Rarity: 2 **Value:** $40.00–60.00 **Reference:** Mk9

Manufacturer: AMT (USA)
Model: Penske/Allison Matador
Rarity: 2 **Value:** $80.00–100.00 **Reference:** Mk10

Four model kits, c. 1970s

Manufacturer: AMT (USA)
Model: Bobby Allison NASCAR Monte Carlo
Rarity: 3 **Value:** $125.00–150.00 **Reference:** Mk11

Manufacturer: AMT (USA)
Model: Bobby Allison Chevy Malibu
Rarity: 3 **Value:** $100.00–125.00 **Reference:** Mk12

Manufacturer: Monogram/Mattel (USA)
Model: T. McEwen's Hot Wheels Mongoose Plymouth Duster
Rarity: 3 **Value:** $80.00–100.00 **Reference:** Mk13

Manufacturer: Monogram/Mattel (USA)
Model: D. Prudhomme's Hot Wheels Snake Plymouth 'Cuda
Rarity: 3 **Value:** $80.00–100.00 **Reference:** Mk14

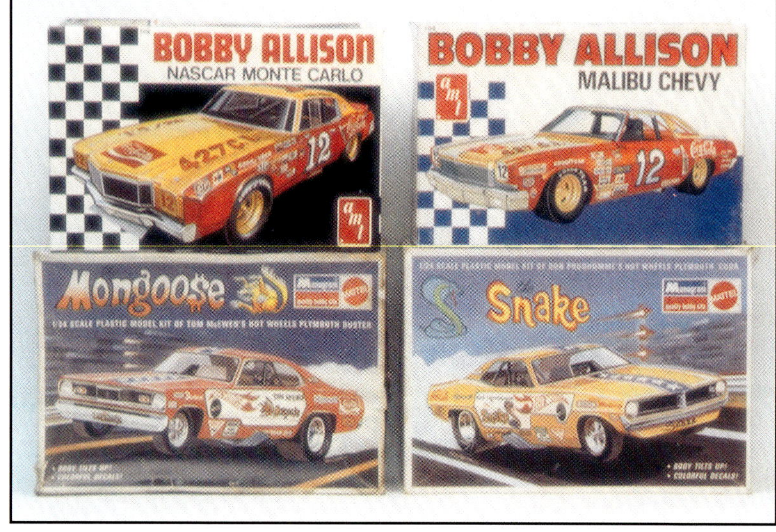

Model kits shown with constructed race cars – Mongoose and Snake.
Reference: Mk13, Mk14

FANTASY

Fantasy

The vehicles pictured in this chapter are fantasies and, as the name implies, are simply the product of someone's imagination. Even though these toys were not licensed by the Coca-Cola Company, it is our firm belief that in most cases their creators did not seek to deceive collectors, but were merely expressing their genuine affection for Coca-Cola. J. Harold Ford, with his wonderful repertoire of masterfully crafted Coca-Cola toys, is a prime example of this. (See Harold Ford, page 218.) Another example of this is the delivery truck pictured on page 215, produced by "Pop" Poppenheimer in the late 1970s. This was one of four different wooden models created by Pop for the members of the Cola Clan, now known as the Coca-Cola Collectors Club. The other three models (not pictured) included a wood Buddy "L" copy, a tanker, and a stake truck. We probably all have a few interesting fantasies within our collections. In fact, we could probably fill another book with just these great make-believe vehicles.

Unfortunately, every once in a while a novice collector pays too much for what he believes to be an original toy. The cast iron toys in this section are examples of this. They were mass-produced for the sole purpose of capitalizing on the Coca-Cola trademark. We happen to be quite fond of the cast iron horse and wagon, nevertheless, these toys have deceived many people through the years. No cast iron toys were ever produced for Coca-Cola – period!

Except for the brass filigree trucks on page 217, no truck in this section is authorized or licensed. In most cases, we have decided to give the "one of a kind" truck a N/A value since we have no criteria for comparsion.

Manufacturer: Unknown (Asia) **Model:** Clydesdale drawn wagon **Composition:** Cast iron **Accessories:** 6–8 plastic cases w/removable bottles
Size: 14¼" **Year:** 1980s **Rarity:** 1 **Value:** $35.00–45.00 **Reference:** FA1

Manufacturer: Unknown (Asia)
Model: Horse drawn cart
Composition: Cast iron
Size: 7"
Year: 1980s
Rarity: 1
Value: $8.00–12.00
Reference: FA2

Manufacturer: Unknown (Asia)
Model: Model T
Composition: Cast iron
Accessories: 6–8 plastic cases w/removable boxes
Size: 9"
Year: 1980s
Rarity: 1
Value: $15.00–18.00
Reference: FA3

Manufacturer: Unknown (Asia) **Model:** Old timey **Composition:** Cast iron **Size:** Left to right: 8", 3¾", 6½" **Year:** 1980s
Rarity: 1 **Value:** Each - $5.00–15.00 **Reference:** FA4, FA5, FA6

Manufacturer: Jim Beam (USA)
Model: Decanter
Composition: Ceramic
Size: 15"
Year: 1986
Rarity: 3+
Value: $250.00–350.00
Reference: FA7
Note: These decanters have appeared in several color variations.

Manufacturer: Unknown
Model: Hobo art
Composition: Metal
Size: 25"
Year: 1980s
Rarity: One of a kind
Value: N/A
Reference: FA8

Manufacturer: Unknown
Model: Hobo art
Composition: Wood
Size: 15"
Year: 1980s
Rarity: One of a kind
Value: N/A
Reference: FA9

Manufacturer: Poppenheimer
Model: Delivery truck, Buddy "L" style
Composition: Wood
Accessories: 20 wooden cases
Size: 18"
Year: 1970s
Rarity: 3+
Value: $125.00–150.00
Reference: FA10
Note: There were three more trucks produced in this series, including a wood Buddy "L" copy, a tanker, and a stake truck. This series was produced in very limited quantities.

Manufacturer: Barclay
Model: Snub nose
Composition: Die-cast metal
Size: 2½"
Year: 1950s
Rarity: 3
Value: Each - $35.00–45.00
Reference: Yellow - FA11; Red - FA11a

Manufacturer: Tootsietoy
Model: Bottle van
Composition: Die-cast metal
Size: 4"
Year: 1980s
Rarity: 1
Value: $10.00–12.00
Reference: FA12

Manufacturer: Bandai (Japan) **Model:** Isetta **Composition:** Tin **Size:** 6½" **Year:** 1950s
Rarity: 3 **Value:** $150.00–175.00 **Reference:** FA13

Manufacturer: M. Houche (France) **Model:** "Coke Bug," VW convertible **Composition:** Plastic **Size:** 6¾" **Year:** 1986
Rarity: One of a kind with these graphics **Value:** N/A **Reference:** FA14

Manufacturer: Unknown
Model: Track wheels
Composition: Plastic
Size: 4"
Year: 1950s
Rarity: 2
Value: $10.00–15.00
Reference: FA15

Manufacturer: Renwal
Model: Pick-up
Composition: Plastic
Size: 3¼"
Year: 1950s
Rarity: 2
Value: $8.00–10.00
Reference: FA16

Manufacturer: Duncan Products (USA)
Year: 1975
Model: Vans (1983 Coca Clan Convention)
Rarity: 1
Composition: Ceramic
Value: $10.00–15.00
Size: 2¾"
Reference: FA17

Manufacturer: Unknown (USA)
Model: Salesman's incentives;
(On wooden stand)
Coca-Cola Collector's Club
Composition: Brass filigree
Size: 4¼", 3¼", 3¾"
Year: 1980s
Rarity: 1+
Value: $35.00–45.00; $12.00–15.00;
$20.00–25.00
Reference: Left to right: FA18, FA19,
FA 20

Manufacturer: Havas (France) **Model:** Fantasy Jeep **Composition:** Watercolor illustration **Size:** 17" x 14"
Year: 1953–54 **Rarity:** One of a kind **Value:** $500.00–700.00 **Reference:** FA21

J. Harold Ford

The following group of trucks were all carved from wood by J. Harold Ford, and we are proud to include them in this section of the book. We have always had a special feeling for these wonderful examples of one man's superior craftsmanship. Harold Ford lives in Louisville, Kentucky, and learned his craft from his father, a woodworker for nearly 40 years. Harold Ford has always produced his toys in very small numbers, usually no more than a dozen of each model. Currently, he is working on a 1936 Ford delivery sedan. He tells us he plans to make only one or two. When asked why he's making so few of these trucks when his toys are in such demand, he laughs and says, "The first one is fun, that's where the challenge is … the rest is work!" Some of Ford's rarest trucks can be seen at the Schmidt Coca-Cola Museum in Elizabethtown, Kentucky.

All of J. Harold Ford's handmade trucks are numbered and signed.

Manufacturer: J. Harold Ford
Composition: Wood, metal
Year: 1980s
Value: One of a kind
Model: Tricycle
Size: Life size
Rarity: 5+
Reference: HF1

Manufacturer: J. Harold Ford
Size: 18"
Value: One of a kind
Model: Schmidt
Year: 1987
Reference: HF2
Composition: Wood
Rarity: 5+

Manufacturer: J. Harold Ford
Model: "Metalcraft" (1 of 10)
Composition: Wood
Size: 10¾"
Year: 1986
Rarity: 5
Value: $300.00–400.00
Reference: HF3

Manufacturer: J. Harold Ford
Model: "Metalcraft" (1 of 12)
Composition: Wood
Size: 13½"
Year: 1991
Rarity: 5
Value: $350.00–450.00
Reference: HF4

Manufacturer: J. Harold Ford
Model: Left - Card truck (1 of 20); Right - Mini 6-pack (1 of 8)
Composition: Wood **Size:** 6½" **Year:** 1987 **Rarity:** 5
Value: Left - $125.00–150.00; Right - $150.00–200.00
Reference: Left - HF5; right - HF6

Truck is numbered in the license plate. This one is #10 of 20 made.

Manufacturer: J. Harold Ford
Model: Case truck (1 of 12)
Composition: Wood from crates
Size: 18¾"
Year: 1989
Rarity: 5
Value: $250.00–350.00
Reference: HF7

TRAINS AND TROLLEYS

Trains and Trolleys

Although this is a book primarily dedicated to toy trucks, we felt that it was appropriate to include a few of their famous wheeled "cousins" – Coca-Cola toy trains and trolleys. We have chosen to include a few of the more popular trains and trolleys in this chapter, including the West German Technofix, the Arnold, and a rare monorail of unknown manufacturing, as well as a colorful Japanese tin lithographed model.

We are also very proud to include one of the most desirable toys ever produced with the Coca-Cola trademark – the American Flyer Train. Here is a toy that has mystified both train and Coca-Cola collectors for years. We are thankful to Randy Schaeffer and Bill Bateman, *The C.C. Tray-ders,* for providing us with photographs and the following brief overview of this rare and magnificent toy.

"The complete train set consists of a locomotive, a tender, two passenger cars, one baggage car, and eight pieces of track. The locomotive is made of cast metal with a spring-driven wind-up mechanism. The tender and cars are made of tin, colorfully lithographed in red, green, yellow, black, and white.

Very little is definitely known about the history of this train. The slogan "Pure as Sunlight" printed on the top of each car was used to advertise Coca-Cola in the late 1920s and early 1930s. Because this Coca-Cola train was not part of American Flyer's commercially available

Manufacturer: American Flyer (USA)
Model: Wind-up train (complete set)
Year: 1928–30
Value: $4,500.00–6,500.00
Composition: Tin
Rarity: 5
Reference: TT1
Note: Set includes locomotive, tender, two passenger cars, baggage car, and eight pieces of track.

line of toy trains, it does not appear in any American Flyer catalogs from the time. However, American Flyer collectors agree the style of the locomotive and cars does indeed date from the late 1920s – early 1930s time period.

From the scant evidence available, it appears that this train was specially manufactured at the request of the St. Louis Coca-Cola bottler. As was the practice among many Coca-Cola bottlers at the time, the St. Louis bottler is known to have used all kinds of items as premiums. In return for handing in a certain number of used bottle caps, the customer could select a free gift. There are vintage photos of the St. Louis bottler's premium room showing the kinds of items available – scooters, wagons, dolls, baseballs and bats, toy stoves, dishes, flashlights, clocks, etc. All were suitably marked with the Coca-Cola logo and carry the "Pure as Sunlight" slogan. Unfortunately, there are no toy trains visible in the existing photographs.

However, there are several reasons to believe that the St. Louis bottler did indeed order the American Flyer train for this use. First, it is not a coincidence that most of the premiums were toys. Children were good collectors of bottle caps and they encouraged their parents to buy Coca-Cola. What better toy than a train? Second, American Flyer's headquarters were in nearby Chicago. Third, and perhaps most convincing, many of the few existing examples of the train were found in the St. Louis area.

It is also possible that the St. Louis bottler made the train available to other bottlers throughout the country. For example, a 1930 premium list from the Los Angeles bottler includes a 'mechanical train and track' available for turning in 600 bottle caps. Of course, it is not known if this is the same train or not, but it seems possible.

Regardless of when it was made, for what reason it was produced, and how it was distributed, the American Flyer Coca-Cola toy train is highly prized. Very few complete examples have survived for today's collectors. Accordingly, the American Flyer Coca-Cola train is considered extremely rare."

Manufacturer: American Flyer (USA)
Model: Passenger car
Composition: Tin
Size: S gauge
Year: 1928–30
Rarity: 5
Value: $1,450.00–1,650.00
Reference: TT2
Note: Close-up of logo at right.

Manufacturer: Technofix (W. Germany) **Model:** #285 Bus Terminal w/wind-up bus **Composition:** Tin **Size:** 40" (opened)
Year: 1950s **Rarity:** 3 **Value:** $350.00–450.00 **Reference:** TT3
Note: There is also a model of this bus terminal with a non-removable key.

Manufacturer: Arnold Spielwaren (W. Germany) **Model:** Wind-up **Composition:** Tin **Size:** 14" **Year:** 1956
Rarity: 4 **Value:** $400.00–600.00 **Reference:** TT4 **Note:** Box variation of TT4 (below)

Manufacturer: Unknown (Japan) **Model:** Freight train, wind-up **Composition:** Tin **Size:** 16½" (Each car - 2½") **Year:** 1960s
Rarity: 4 **Value:** $400.00–500.00 **Reference:** TT5

Manufacturer: Unknown (W. Germany)
Model: Monorail, wind-up
Composition: Tin
Size: Oval track - 23" diameter; car - 5"
Year: 1950s
Rarity: 5
Value: $400.00–500.00
Reference: TT6
Note: Detail of monorail track and Coca-Cola sign.

Manufacturer: Payva (Spain)
Model: Tranvia, friction power
Composition: Plastic
Size: 7"
Year: 1970s
Rarity: 3
Value: $125.00–175.00
Reference: TT7

Manufacturer: Diamond Toy Makers (Hong Kong)
Model: Jolly Trolley
Composition: Plastic
Size: 3½"
Year: 1970s
Rarity: 1
Value: $10.00–15.00
Reference: TT8

Manufacturer: NBD (Hong Kong)
Model: Coca-Cola circus train w/spinning top, wind-up
Composition: Plastic
Size: 4¼"
Year: 1972
Rarity: 4
Value: 40.00–60.00
Reference: TT9
Note: 1,500 produced
(Also came with yellow top - same value)

Manufacturer: Lionel (USA) **Model:** 027 gauge **Composition:** Plastic, metal **Size:** 5 cars plus track **Year:** 1973 **Rarity:** 2
Value: MIB - $350.00–450.00 **Reference:** TT10
Note: There is also a very rare Chattanooga 75th anniversary version, c. 1974. **Rarity:** 5 **Value:** $650.00–850.00

Manufacturer: Model Power (Yugoslavia) **Model:** Coca-Cola **Composition:** Plastic **Size:** HO scale **Year:** 1980s
Rarity: 3 **Value:** Each set - $100.00–125.00 **Reference:** TT11

Manufacturer: Lima Models (Italy) **Model:** Boxcar **Composition:** Plastic **Size:** N scale **Year:** 1980s
Rarity: 1 **Value:** $15.00–20.00 **Reference:** TT12

Manufacturer: Lima Models (Italy)
Model: Boxcar
Composition: Plastic
Size: HO scale
Year: 1980s
Rarity: 1
Value: $15.00–20.00
Reference: TT13

Manufacturer: Lima Models (Italy)
Model: Boxcar
Composition: Plastic
Size: HO scale
Year: 1980s
Rarity: 1
Value: $15.00–20.00
Reference: TT14

Manufacturer: Lima Models (Italy) **Model:** Lima Crick (boxed set of 4) **Accessories:** Cut-out landscape on box
Composition: Plastic **Year:** 1980s **Rarity:** 2 **Value:** Set - $60.00–80.00
Reference: TT15

Manufacturer: Liliput (Austria)
Model: Boxcar
Composition: Plastic
Size: HO scale
Year: 1980s
Rarity: 1
Value: $15.00–20.00
Reference: TT16

Manufacturer: Athearn (USA)
Model: Boxcar
Composition: Plastic
Size: HO scale
Year: 1970s
Rarity: 2
Value: $15.00–20.00
Reference: TT17

Manufacturer: Atlas Car/AkSarBen Hobby (USA) **Model:** Limited collector's overrun #8612 **Composition:** Plastic **Size:** N scale
Year: 1970s **Rarity:** 3 **Value:** $25.00–35.00 **Reference:** TT18

Manufacturer: Tyco (USA) **Model:** Boxcar **Composition:** Plastic **Size:** HO scale **Year:** 1970s
Rarity: 3 **Value:** $25.00–35.00 **Reference:** TT19

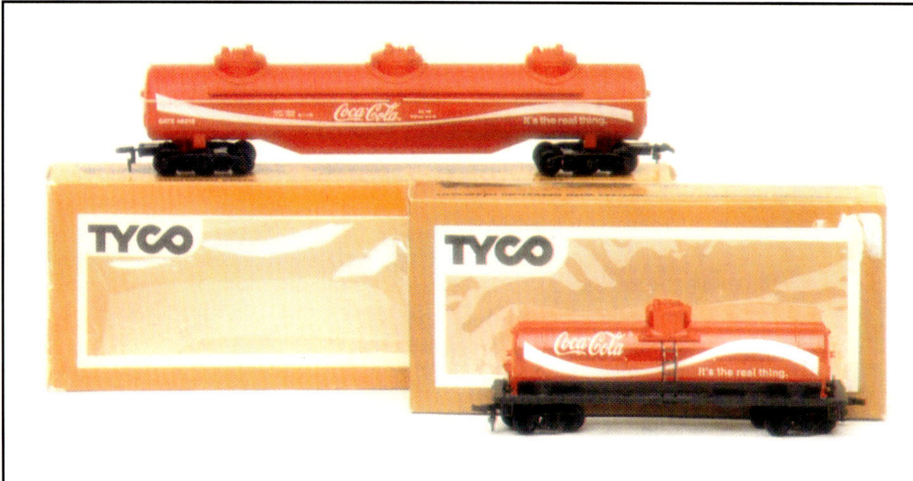

Manufacturer: Tyco (USA)
Model: Tankers
Composition: Plastic
Size: HO scale
Year: 1980s
Rarity: 1
Value: Each - $10.00–15.00
Reference: TT20, TT21

Manufacturer: Tyco (USA) **Model:** Electric Trucking **Accessories:** 2 trailers
Composition: Plastic **Size:** HO scale **Year:** 1980s
Rarity: 1 **Value:** $10.00–15.00 **Reference:** TT22

Manufacturer: Markatron Inc. (USA) **Model:** Limited Express **Composition:** Plastic **Size:** 027 gauge **Year:** 1986
Rarity: 1 **Value:** $125.00–150.00 **Reference:** TT23

Manufacturer: Markatron Inc. (USA) **Model:** Limited Express 2 **Composition:** Plastic **Size:** 027 gauge **Year:** 1987
Rarity: 1 **Value:** $125.00–150.00 **Reference:** TT24

MISCELLANEOUS

Safe Drivers Award brass belt buckle, c. 1970s. **Rarity:** 3 **Value:** Each - $20.00–25.00

Safe Drivers Award copper or silver pins, c. 1940s.
Rarity: 4 **Value:** Each - $50.00–60.00; Set of 5 - $250.00

Miscellaneous Coca-Cola memorabilia.

RESOURCES

Buffaloe, Bob. "Metalcraft: The Toy," *The Cola Call*, March 1981.

Caldwell, Dave and Melba. "Even Scrooge Would Have Loved Toy Coke Trucks," *The Cola Call,* November 1982.

"Children in 1 of Every 5 Montreal Homes Now Play With Miniature Trucks," *The Red Barrel,* Vol. XII, No. 5, May 15, 1933.

Colpitts, Harold L. Lledo Coca-Cola Models.

Gobetti, Jon. "The Man Who Loves Smith-Miller Toys," *Street Rod Action,* October 1993.

McCollough, Albert W. *The New Book of Buddy "L" Toys, Volume I,* Greenberg Publishing Company, Inc., 1991.

McCollough, Albert W. *The New Book of Buddy "L" Toys, Volume II,* Greenberg Publishing Company, Inc., 1991.

Pinsky, Maxine A. *Greenbergs's Guide to Marx Toys, Volume I,* Greenberg Publishing Company, Inc., 1988.

Shartar, Martin and Norman Shavin. *The Wonderful World of Coca-Cola,* Capricorn Corporation, Inc., 1981.

West, John. "Remembering Al Korte (Part 2)," *Antique Toy World.*

EXPERT SOURCES

Bill Bateman and Randy Schaeffer – American Flyer Coca-Cola train

Fred Thompson – Smith-Miller

Richard McNary – Marx Toys

Lyle Dever – toy trucks in general

Allan Petretti – Coca-Cola collectibles

Ugo Fadini – European toy trucks

INDEX

Schroeder's

ANTIQUES

Price Guide

. . . is the #1 best-selling antiques & collectibles value guide on the market today, and here's why . . .

8½ x 11, 608 Pages, $12.95

• *More than 300 advisors, well-known dealers, and top-notch collectors work together with our editors to bring you accurate information regarding pricing and identification.*

• *More than 45,000 items in almost 500 categories are listed along with hundreds of sharp original photos that illustrate not only the rare and unusual, but the common, popular collectibles as well.*

• *Each large close-up shot shows important details clearly. Every subject is represented with histories and background information, a feature not found in any of our competitors' publications.*

• *Our editors keep abreast of newly developing trends, often adding several new categories a year as the need arises.*

If it merits the interest of today's collector, you'll find it in *Schroeder's*. And you can feel confident that the information we publish is up to date and accurate. Our advisors thoroughly check each category to spot inconsistencies, listings that may not be entirely reflective of market dealings, and lines too vague to be of merit. Only the best of the lot remains for publication.

Without doubt, you'll find
SCHROEDER'S ANTIQUES PRICE GUIDE
the only one to buy for
reliable information and values.

COLLECTOR BOOKS
A Division of Schroeder Publishing Co., Inc.